Expedition Field Techniq

BIRD SURVEYS

Colin Bibby, Martin Jones and Stuart Marsden

Published by the Expedition Advisory Centre
Royal Geographical Society
(with The Institute of British Geographers)
1 Kensington Gore
London SW7 2AR

Tel +44 (0)171 591 3030   Fax +44 (0)171 591 3031
Email eac@rgs.org   Website www.rgs.org

October 1998

ISBN 0-907649-79-3

Front Cover: Chinese Grouse *Bonasa sewerzowi*, near-threatened and endemic to the Himalayan foothills of Western China. Artist: Richard Thewlis.

Expedition Field Techniques
BIRD SURVEYS

CONTENTS

Acknowledgements	1
Foreword	2
Introduction	3

Section One: Why Count Birds?
1.1	Species	5
1.2	Sites	7
1.3	Important Bird Areas	8
1.4	Habitats	10
1.5	Choice of methods	11

Section Two: Study Design
2.1	Introduction		15
	2.1.1	Total counts	15
	2.1.2	Sampling and bias	15
	2.1.3	Sampling, precision and accuracy	16
	2.1.4	Relative and absolute estimates	17
	2.1.5	Measuring and increasing precision	18
2.2	Study Design		19
	2.2.1	Choosing the right time and conditions	19
	2.2.2	Observer bias	21
	2.2.3	Sample sizes and replication	24
	2.2.4	Positioning your sampling effort	26
2.3	Pilot surveys and training		29
2.4	Safety, team size and logistics		31
	2.4.1	Health and safety	31
	2.4.2	Logistics	31
	2.4.3	What size team?	31
2.5	Which methods to use		32
	2.5.1	Introduction	32
	2.5.2	Bird densities	33
	2.5.3	Point counts and line transects	34

Section Three: Estimating bird densities using distance sampling
3.1 Introduction   35
3.2 Distance sampling using line transects   36
     3.2.1 Introduction   36
     3.2.2 Positioning of transects   36
     3.2.3 How many transects and how long should they be? 37
     3.2.4 Collection of data   37
     3.2.5 Double counting   40
     3.2.6 Variable distances and distance bands   40
     3.2.7 Distance estimations to groups   41
     3.2.8 Sample sizes   41
3.3 Distance sampling using point counts   41
     3.3.1 Introduction   41
     3.3.2 Location of census stations   42
     3.3.3 How many census stations?   43
     3.3.4 How long should the count period be?   43
     3.3.5 Data collection   43
     3.3.6 Variable distances and distance bands   44
     3.3.7 Sample sizes   45
3.4 Examining the data   45
     3.4.1 Bird detectability and the detection curve   45
     3.4.2 Shape criteria   47
3.5 Using the DISTANCE software   50
     3.5.1 The basic model   50
     3.5.2 Alternative models   50
     3.5.3 Testing model fit   51
     3.5.4 Inputting data   51
     3.5.5 Understanding the output   52

Section Four: Counting single-species
4.1 Is a single-species study appropriate?   53
4.2 Narrowing down the search: information is the key   53
4.3 The nature of the beast: rarity   57
4.4 Problem species and difficult habitats   60
     4.4.1 Bird colonies/aggregations   60
     4.4.2 Cryptic and understorey birds   61
     4.4.3 Canopy species   61
     4.4.4 Mixed-species flocks   62
     4.4.5 'Aerial' birds   62
     4.4.6 Nocturnal and crepuscular birds   62
     4.4.7 Other 'problem' birds (waterbirds,
            birds of prey, migrants, etc.)   63

		4.4.8 Special habitats/niches	63
4.5		Tailoring distance sampling methods for individual situations	64
4.6		Specific techniques for special cases	67
	4.6.1	Spot mapping	67
	4.6.2	Counting birds at or near aggregations	68
	4.6.3	'Look down' methods from vantage points	70
	4.6.4	Nest searching/counting limited resources	71
	4.6.5	Interviews with local people	73
	4.6.6	Mark-recapture/banding/home ranges	73
	4.6.7	Scientific birding	74
4.7		Interpreting and presenting results of specific studies	75

Section Five: Assessment of sites: measurement of species richness and diversity

5.1		Introduction	76
5.2		Compiling a species list	76
5.3		Standardising recording methods	81
	5.3.1	Species discovery curves	81
	5.3.2	Encounter rates	83
	5.3.3	Mackinnon lists	84
	5.3.4	Timed species-counts (TSCs)	86
	5.3.5	Mist-netting	86
5.4		Analysis of data	87
	5.4.1	Predicting total number of species from species discovery curves	87
	5.4.2	Analysis of encounter rate data	89
	5.4.3	Analysis of Mackinnon list data	91
	5.4.4	Analysis of TSC data	91
	5.4.5	Analysis of mist-net data	92
5.5		Discussion	94
5.6		Sources of information for the recording of bird sounds	97

Section Six: Bird-habitat studies

6.1		Why study habitats?	99
6.2		Broad scale habitat studies	100
	6.2.1	What features to map?	100
	6.2.2	Sources of data for mapping	101
	6.2.3	Verification of a map and sampling	102
6.3		Fine scale bird-habitat studies	102
	6.3.1	Different approaches/survey designs	102
	6.3.2	What habitat features to record and how?	103

| | | 6.3.3 | Preparation of data for analysis | 108 |

6.4 Analytical approaches 110
    6.4.1 Summary statistics 110
    6.4.2 Indices 110
    6.4.3 Graphical and linear regression approaches 111
    6.4.4 Logistic regression 113
    6.4.5 Reducing the dimensions 113
    6.4.6 Interpreting and testing the results 114

Section Seven: Maximising the impact of the work
7.1 Basic communications 115
7.2 Culture, politics and diplomacy 116
7.3 Summary reporting 118
7.4 Scientific reporting 122
7.5 Archiving the data 124

Section Eight: References and further reading 125

**Section Nine: Sample input and output
                  files for DISTANCE Program** 127

The BP Conservation Programme 134

Acknowledgements

This book has been assembled as a collaborative effort. Thanks are due to authors of individual sections who additionally contributed to discussions about the whole project and on sections other than their own.

We are particularly grateful to the following who read and helpfully criticised large parts of the text, bringing to bear their experience from around the world: Rod Hay, Sam Kanyamibwa, Borja Mila, Michael Poulsen, Richard Thomas and Hazell Thompson. Individual sections have benefited from the input of the following: Leon Bennun, S. Buckland (for technical help with DISTANCE), Paul Dolman, Bill Sutherland, David Whitacre and Rob Williams.

Ian Burfield, John Pilgrim and Rob Williams helped in many ways in co-ordinating and bringing the project to fruition. Katharine Gotto from the BP Conservation Programme helped and encouraged throughout. We appreciate the help of Shane Winser and Timothy Jones at the Expedition Advisory Centre of the Royal Geographical Society for overseeing the final production process. Financial support from BP has allowed us to keep down the cost of the publication so that it can be distributed more widely.

Author Profiles:

Colin Bibby is the director of Science and Policy at BirdLife International's UK headquarters. His interests are in the collection and use of scientific information for the conservation of birds and their habitats.

Martin Jones is a senior lecturer in the Behavioural and Environmental Biology Group in the Department of Biological Sciences at Manchester Metropolitan University. His research interests are the ecology and conservation of tropical birds and the effects of habitat change on island birds and butterflies.

Stuart Marsden is a lecturer in Behavioural and Environmental Biology in the Department of Biological Sciences at Manchester Metropolitan University. His main research interests are in the ecology and conservation of parrots and other tropical birds.

*Figure 24 has been kindly reproduced with the permission of Oxford University Press. Taken from: Mackinnon, J. and Phillips, K. (1993) *A Field Guide to the Birds of Sumatra, Java and Bali.*

FOREWORD

This book is a vital tool for everyone wishing to contribute to our knowledge of the world's birds and to bird conservation. Effective conservation planning can only be based on a sound knowledge of the species, sites and habitats in need of protection.

Despite birds being the best known class of living organisms there are still substantial gaps in our knowledge of the distributions, abundances and densities of species. Birds have been demonstrated to serve as good indicators of biodiversity and environmental change and as such can be used to make strategic conservation planning decisions for the wider environment.

BirdLife International is delighted to have been able to collaborate with the Expedition Advisory Centre of the Royal Geographical Society (with the Institute of British Geographers) to produce this much needed volume. The editors and authors are all experienced in their subjects and the book has been reviewed and refined by specialists from around the world.

To make the best decisions, it is most important that the information on which such decisions are made is as accurate, systematic and representative as possible. The methods in this book will enable the user to survey birds simply and effectively.

This book will be much used by conservationists, researchers and birders, both amateur and professional throughout the world. I expect it will make a significant contribution towards the furthering of knowledge about the avifauna of the world and towards safeguarding biodiversity.

Dr Michael Rands
Director and Chief Executive, BirdLife International

INTRODUCTION

There are many reasons for counting birds and a large and rather forbidding literature on the subject. Birds are among the best known parts of the Earth's biodiversity. But nevertheless soundly quantified knowledge is far from complete for most species and regions. We believe that this is an obstacle to conservation of birds which ornithologists can help to rectify. Birds are relatively easier to count than most other wildlife and ornithologists have a distinct contribution to make to biodiversity conservation by improving our understanding of the planet, the location of biodiversity and threats it faces from non-sustainable practices.

Unashamedly we have focused strongly on the application of counting methods for conservation. For this reason, we have to a degree biased coverage to forests and the tropics where so much biodiversity resides. We have tried to impart general principles and some practical techniques in a clear and simple manner. We may be criticised for citing few references but have done this to help readers get going without feeling that there is a challenge to read a huge literature first. We hope that we have illustrated enough of the principles behind bird counting to enable the reader to take a critical attitude to their planned study. Many of the principles are common to any method applied to any species or habitat. There are other important kinds of study of birds, such as ecology or population dynamics which we have not covered at all.

This book is intended to help conservation professionals or students plan field surveys at home or abroad. It is not possible to count birds without a good field knowledge and ability to identify them, but this aside, we have tried to make no prior assumptions about the skill of the reader. We have tried to write both for internationally travelling students who have contributed so much in the past and for biologists in developing countries who have so much more to contribute in the future.

We will judge the success of this book by the extent to which we see it cited in studies of important areas or threatened species which come through to influence conservation decisions at local and national level around the world. Ultimate success will be judged by the impact that you, the reader, can have for conservation by collecting new and important data and supporting the development of more effective conservation work wherever you live.

Section 1
WHY COUNT BIRDS?

Colin Bibby

There are many good reasons for counting birds but this guide aims to promote better knowledge to help conservation. A recurring theme will be that well designed field studies start with a clear purpose. The sharp definition of purpose is probably one of the more difficult steps in designing a good study. Once a purpose is clear, it becomes much more obvious whether any particular study design has a reasonable chance of working and whether there are variants which would be better.

Most surveys target a particular species or a particular place. Important questions may arise about the use a species makes of habitats or, at a site, the condition of different habitats and the species that occur in them. The division between species, sites and habitats will recur throughout the book.

1.1 Species

There is an urgent need to know more about the world's most threatened bird species. These are officially listed in the BirdLife International publication, *Birds to Watch 2* (Collar *et al.* 1994). The definition of globally threatened species has been agreed by the Species Survival Commission of the World Conservation Union (IUCN, 1994). The system puts different species into categories according to a set of criteria (Figure 1). The most important data are population size and range, and trends in one or other of these. Trends are impossible to measure unless some baseline has previously been set. For most species this has yet to be done. In addition, threat might be measured as known or inferred change of extent or condition of habitat. This can only be applied if the habitat requirements of the target species are reasonably well known. For the majority of bird species, and especially for many threatened species, these most basic parameters are simply unknown. By 1992 for instance, less than a quarter of threatened species in the Neotropics had been subject to any formal counting.

One way of telling how effective this book has been will be the rate at which successive editions of *Birds to Watch* show development of quantitative knowledge on threatened species. Viewed the other way round, perusal of *Birds to Watch* offers anyone who is interested a clear challenge to get out into the field and collect some new data of real value. Such information will not only help to ensure that threatened species are correctly recognised, but will also help planning for their conservation. *Birds to Watch* also has a category of near-threatened. This covers species given a precautionary listing until sufficient data have been collected and analysed to decide how their conservation status should be described.

Figure 1. Some of the criteria used for identification of IUCN Red List Categories (from IUCN, 1994).

	Critical	Endangered	Vulnerable
Population decline	>80% in 10 yrs	>50% in 10 yrs	>20% in 10 yrs
Extent of occurrence *	<100 km^2	<5000 km^2	<20,000 km^2
Area of occupancy *	<10 km^2	<500 km^2	<2,000 km^2
Population level *	<250 individuals	<2,500 individuals	<10,000 individuals
Population level	<50 individuals	<250 individuals	<1,000 individuals

* These criteria do not categorise species alone but have to be met in combination with other factors indicating declines, or fragmentation of populations. Data may be known, estimated, inferred or suspected but need to be documented. Extent of occurrence is overall range size – the area of a polygon embracing localities. Area of occupancy is the total area of habitats occupied, so may be much smaller but depends on a knowledge of habitat preferences and extent of suitable habitat.

Many countries have official lists of species of national priority. These are often based on similar ideas about range, numbers and trends but with lower thresholds. Other species attract attention because they may be potential environmental indicators, or simply because they are popular – so called 'flagship species'. Given the pressing need for information on globally threatened species, we would urge species-oriented work in remote places to concentrate on these. There is merit in collecting quantitative data on as many species as possible at the same time. This is often a sensible approach because looking for threatened birds can be time consuming with rather little data in return. Better then, to collect some systematic information on other species while looking for your threatened 'target' species. In other circumstances, good data can often be collected by focusing on a single-species especially if using a technique like play-back.

1.2 Sites

Species may be common currency to biologists but they are not very practical targets for conservation. Put simply, there are just too many to be treated one by one – it is important to remember that most species are not birds and that the vast majority are not even known to science. A more practical unit for conservation is the protected area. This might be strictly protected for nature conservation or, particularly in the developing world, might include utilisation by humans.

BirdLife International has demonstrated (ICBP, 1992) the location of 218 Endemic Bird Areas (EBAs) to which one quarter of bird species on Earth are confined (Figure 2). The Albertine Rift Mountains in central Africa, for example, are about 56,000 km^2 in extent and have 36 endemic bird species. These EBAs, which occupy just 5% of the Earth's land surface, embrace some three quarters of all threatened species. They are therefore critical regions for conservation. Within the EBAs, there is a pressing need to narrow down relatively large regions into sites of a size that are already protected or may become so in the future. Full documentation of the EBAs is available in Stattersfield *et al.* (1998). This adds a geographic dimension to the inspiration available from *Birds to Watch 2* and indicates many areas in need of ornithological exploration.

To contribute helpfully to the conservation of sites, one needs to know where they are and what occurs within them. In remote areas with poor access, even locating the boundaries of a site may be practically (and conceptually) very difficult. Well designed fieldwork needs to be clear about its geographic boundaries.

Pioneering new areas is obviously exciting but there are also large gaps in our knowledge of the birds of existing protected areas. Filling these can have practical use to local managers and conservationists. The information helps us to understand which species are most important, which might require special management because of their poor status and which might be so rare as to be in need of further protection elsewhere within their ranges. Baseline counts will come to be greatly valued when repeated in the future. They will show which species have declined in numbers and hence need additional management if they are not to disappear.

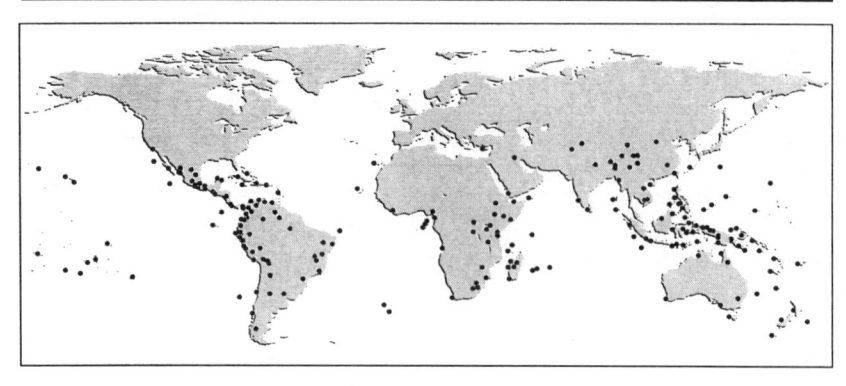

Figure 2. Location of 218 Endemic Bird Areas (taken from ICBP, 1992; Stattersfield et al. 1998). Each of these areas has at least two bird species solely confined to it. In total they occupy about 5% of the Earth's land surface but one quarter of all bird species and about three quarters of all threatened species are confined to them. As a result, these areas are of very high priority for further exploration and conservation.

1.3 Important Bird Areas

BirdLife International is in the process of documenting sites of global importance for conservation in its Important Bird Areas (IBA) programme (Figure 3).

Figure 3. BirdLife International Important Bird Areas programme.

IBAs :
• are places of international significance for the conservation of birds at the global, regional or sub-regional level;
• are practical tools for conservation;
• are chosen using standardised, agreed criteria applied with common sense;
• must, wherever possible, be large enough to support self-sustaining populations of those species for which they are important;
• must be amenable to conservation and, as far as possible, be delimitable from surrounding areas;
• will preferably include, where appropriate, existing protected areas;
• should form part of a wider, integrated approach to conservation that embraces sites, species and habitats.

The function of this programme is to identify and protect a network of sites, at a biogeographic scale, that are critical for the long-term viability of naturally occurring bird populations, across the range of those bird species for which a sites-based approach is appropriate. Sites in Europe (Grimmett and Jones, 1989) and the Middle East (Evans, 1994) have already been documented and work is developing in the rest of the world. IBAs are identified by a set of globally-agreed criteria. This is important to ensure the credibility of the whole set – a site cannot just be an IBA because someone feels it is important; there has to be some supporting data. Criteria fall into four groups (see Figure 4):

- globally threatened species
- restricted-range species
- biome restricted assemblages
- congregatory species

Population estimates are required to identify IBAs for globally threatened species and for congregatory species (generally waterfowl or seabirds) and appropriate techniques are discussed in Sections 3 and 4. Restricted range species and assemblages may be sufficiently measured by one of the methods described in Section 5. Biomes and their definitive birds have been described for the IBA programmes for the major continents and more information on the IBA programme is available from the partner organisations or offices of BirdLife.

Figure 4. Criteria for globally Important Bird Areas (part).

Category	Criterion
A1 Globally threatened species	The site regularly holds significant numbers of a globally threatened species, or other species of global conservation concern.
A2 Restricted-range species	The site is known or thought to hold a significant component of a group of species whose breeding distributions define an Endemic Bird Area (EBA)
A3 Biome-restricted assemblages	The site is known or thought to hold a significant component of the group of species whose distributions are largely or wholly confined to one biome.
A4 Congregations	i) Site known or thought to hold, on a regular basis, $\geq$ 1% of a biogeographic population of a congregatory waterbird species. Or: ii) Site known or thought to hold, on a regular basis, $\geq 1\%$ of the global population of a congregatory seabird or terrestrial species. Or: iii) Site known or thought to hold, on a regular basis, $\geq$ 20,000 waterbirds or $\geq$ 10,000 pairs of seabirds of one or more species. Or: iv) Site known or thought to exceed thresholds set for migratory species at bottleneck sites.

Determining boundaries of IBAs can be tricky. Sites should, as far as possible:

- be different in character or habitat or ornithological importance from the surrounding area;
- exist as an actual or potential protected area, with or without buffer zones, or be an area which can be managed in some way, as a unit, for nature conservation;
- alone or with other sites, be a self-sufficient area which provides all the requirements of the birds (that it is important for) which use it during the time they are present.

Simple, conspicuous boundaries such as roads, rivers, railway lines, etc. may be used to delimit site margins while features such as watersheds and hilltops may help in places where there are no obvious discontinuities in habitat (transitions of vegetation or substrate). Boundaries of ownership may also be relevant.

1.4 Habitats

Within sites, it is fairly evident that habitat is likely to be an important determinant of the distribution and number of birds. For sites which are not protected, habitats might be changing, for instance as a result of logging. Adequate management obviously depends on understanding the relationship between birds and their habitats. If a study is oriented to a particular species, it is also evident that questions about its distribution, ecology and threats to its status, will partly be answered with an understanding of its habitat requirements. While much about a bird's ecology might be studied directly in terms of its diet, foraging behaviour or population dynamics, important knowledge of habitats can be gleaned from good census studies.

Explicit questions about habitats are likely to take a certain form, e.g. what are the major variations of habitats around here, and how does the abundance of birds vary with them? Variations might be of natural origin, for instance by soil type, along a gradient of rainfall or by altitude. Important variations might have human origin, such as the degree of impact of logging on forest structure, from mature to selectively-logged to clear-felled and regrowing secondary stands.

Questions of this kind need a method for recognising and describing variations in habitats. They also call for a well designed study capable of collecting sufficient data across the range of habitats involved. Section 6 has been written to indicate some methods for measuring and describing vegetation and habitats because this is such an important part of answering questions about sites and species.

1.5 Choice of methods

A simple checklist of questions (Figure 5) should help to ensure good design of a study by pointing out problem areas where methods might not be properly linked to the original question.

Figure 5. Eleven questions to answer in designing a study.

- what is the question?
- who will use the results?
- who are the appropriate contacts?
- where are the boundaries of the study?
- how is the effort going to be distributed?
- what methods will be used in the field?
- is the method good enough for the purpose?
- is the study realistic?
- what preparations are needed?
- how will the data be analysed?
- how will the results be disseminated?

What is the question? The more simply one or more questions can be posed the better. What is the status of the regionally endemic bird species in area A? What are the effects of logging on birds in place B? What is the world population of bird C?

Who will use the results? Protected area managers might have a very clear idea of why they want particular information and thus what kind of field data will be needed. Even in the same place, a research study on the population viability of a particular species will need a different approach.

Who are the appropriate contacts? Section 7 elaborates the point that the impact of your study will be greatly influenced by the quality of your local contacts and diplomacy. In preparation you need local points of contact to help plan a study that will be helpful to local or national authorities in a position to use the information you discover. Obviously you need access to the most up to date intelligence on what surveys have recently been done or are planned and where the gaps are. Obvious starting points are BirdLife Partners or offices, other established bird clubs, natural history societies, conservation organisations or universities. Government departments might help if you can match your interests with theirs.

Where are the boundaries of the study? This is a more difficult question than it might seem. The status of a globally threatened species needs to be assessed across its whole range, which might be large and poorly known. Even a single protected area in the tropics might well be too large to be easily covered in one visit. If the study cannot embrace such a large area, then it is important to define the smaller area that it will cover properly. Otherwise, the results might be very difficult for anyone else to use subsequently.

How is the effort going to be distributed? If the boundaries of the study area are too large to allow complete coverage, then the study design must include sampling. Sampling can be a powerful way of inferring general patterns if it is done correctly. If it is ignored, or done badly, it can be very misleading. Sampling might be random, which is good in theory but tough in practice. Stratification may often be appropriate, see Section 2. With different habitats sampled at appropriate levels the effort going into different plots or strata should ideally be planned, but if, as is often the case, this is not realistic, then it at least needs to be measured and documented.

What methods will be used in the field? There are only half a dozen essentially different things that bird counters do in the field. They all require some discipline slightly greater than pure bird-watching, (tempting though this might be in a new place rich in exciting species). The simplest methods (Section 5) add very little more than some basic note taking to bird-watching.

If one or two methods are selected in advance, there will be merit in designing a data recording sheet. This has the advantage of reminding people what data they need to record, promotes standardisation and allows a daily check of how things are going. It is also good for data security because if the previous day's results can be safely stored away, there is no risk of loss when going into the field again with a note-book.

Is the method good enough for the purpose? It is not easy to describe, for all circumstances, how this test might be passed or failed. As a rule of thumb, you need a minimum of about ten records of a species to make a reasonable estimate of its abundance. To describe a forest bird community or the birds of a habitat type requires a minimum of about 50 point counts or 10km of transect (see Section 3). With less formal methods, this might be 10–20 species lists, or one hour lists depending on the richness of the habitat (see Section 5). An ideal study would not only cover several habitat types but would also provide two to four replications of each in order to see whether or not there is consistency in the generality of the results.

Unappreciated bias is a problem with the most devastating potential to make the results of a study useless. If a key part of the range of habitats in an area has not been sampled then it is not possible to make any inferences as to what might be there. If this is appreciated it is no problem – the results of the study simply apply to a smaller known area. If it is not appreciated, then it is not possible to say how applicable the results are, or to what area. There are many other causes of bias (Section 2) which can be problematic if left to themselves. What happens if the observer who is not actually a very experienced ornithologist is the only person collecting data in one habitat type, while a very observant and experienced recorder gathers the data in another? What happens if key elements of the method are allowed to vary in just one part of the study?

There is a common tendency to believe that any results other than the very precise are not much use. It can be difficult to get precise results, particularly in circumstances where fieldwork is arduous. This does not mean that it is difficult to get any worthwhile results from such places. For many possible questions, even quite imprecise data are enough and certainly much better than no data at all. The handling of bias and questions about precision are discussed in Section 2.

What preparations are needed? Time spent in preparation is rarely wasted. Have you planned what you need to prepare for? Have you got all the relevant background information on the area and its birds? How are you going to learn to recognise birds or habitats in the area? Are there any key people who might be able to help? You really need to talk with or involve people who have used the planned methods before and who know about the study area. Would vegetation maps, air photos or satellite images help? Where will you find them?

For a trip to an unfamiliar foreign country, it might take most of the preceding year to prepare fully. Even on a project to a remote part of one's own country, preparations might take several months. Having arrived on site, it is a good idea to practice and check methods before getting going. This particularly includes identification of birds or trees (if needed) and measures of distances. The better the prior preparation, the quicker this stage will be.

Is the study realistic? It is very common to be over-ambitious in designing a study. It may actually be more useful to set and achieve a modest objective than to half accomplish something grander and end up with an unfinished job of limited value to anyone. As broad guidance, consider that arranging permissions and local diplomacy might take as much as two weeks. Even when out of the city, expect to use about half your days for local travel, domestic maintenance or illness. Depending on where you are and the season, days will sometimes be lost to the weather. In a new area it might take two weeks to become familiar with the birds and design the practical details of a study. So an eight week trip to a remote area might allow as little as 15 days of fieldwork. During those days, one person can realistically make ten point counts or walk a 4km transect per day, though in ideal circumstances these figures might be doubled. These approximate limits are set by confining fieldwork to the best time of day, by energy consumed in access and by the rate at which it is possible to sustain fieldwork day after day. For safety reasons it might be better to work in pairs and you might need people guarding the camp or running domestic chores.

How will the data be analysed? The benefit of thinking about analysis before collecting any data is that it reduces the likelihood of anything important slipping past unmeasured or unconsidered. Are there any considerations to make numerical data available for computer entry and analysis? Is all the field data properly coded for location, altitude and habitat measurements? Does anyone know how to use the software for estimating densities?

How will the results be disseminated? There is no point in analysing the data if the results are not going to be communicated to somebody (see Section 7). The more carefully you think in advance about what your report will be like, the more likely that you will end up collecting appropriate data in a suitable way. Another important thought is whether the study will be repeated by yourselves or someone else. If it is important, one would certainly hope so. Would it be possible for someone else to be able to repeat what you plan to do?

If you can give clear answers to these questions your study deserves to work and it is time for you to go into the field. The clarity of your prior thinking will repay you well. Indeed, it might even pay you in advance since evidence of careful planning and prior thought is very attractive to funders. In truth, realities in the field will intrude on the best laid plans and you will need the flexibility to change things as you learn more.

Section 2
STUDY DESIGN

Martin Jones

2.1 Introduction

The best studies are the ones where the participants not only have a very clear idea of their aims but also understand the methods they are going to adopt and – crucially – know how they are going to analyse the data. Once your aims have been formulated, and with the help of this book, you should be able to identify the appropriate methods and analysis for your study. At least one team member should then become fully conversant with all aspects of data analysis techniques before field work commences. Prior to beginning fieldwork it is possible to plan the study in broad terms, but fine tuning will always depend upon local knowledge, results of pilot studies and initial analysis of the results as they come in. If it is apparent that the aims of the study will not be adequately met you have two options – either redesign the work or modify the aims!

2.1.1 Total counts
In a few cases, it may be possible to make a total and accurate count of a species, either within its world range or within a defined habitat or protected area. A more likely situation is that total counts are impossible and some sort of sampling is required. Sampling is always needed for establishing habitat associations, for multi-species surveys, and for diversity studies.

2.1.2 Sampling and bias
The basic idea which underpins sampling is that because we cannot count a whole population or bird community, we take samples and extrapolate our results to provide estimates of the true population sizes or species diversities. In the same way, we might sample a variety of habitats to try to build up a true picture of what a species' habitat requirements really are. The problem with any sort of sampling is that there are many ways in which the sampling regime could be biased. For example, many birds are more active and vocal early in the morning, so if two forest areas are censused, one between 0600 and 0800h and the other between 1300 and 1500h, the results cannot be compared; the first area may seem to have more birds but is this because of a real difference in the bird populations, or just because the birds were easier to see and hear? The sampling regime was obviously biased, and there are many other ways in which bias can affect the outcome of any bird counting exercise. Another example of bias is comparing results from a noisy

environment (e.g. riparian forest) with a 'quiet' habitat. Understanding the causes of bias and dealing with it in the appropriate way is the most important part of study design and is dealt with in Section 2.2.

2.1.3 Sampling, precision and accuracy

If we are estimating a population, assessing species diversity or studying habitat associations, we would never usually take just one sample. Even if we could eliminate all sources of sampling bias, natural variation in habitats and bird distribution will mean that samples are different. The term 'precision' describes the closeness of repeated sample estimates to each other, while 'accuracy' describes how close the estimates are to the real value.

For example, if we wanted to estimate the population density of a particular bird species in an area of forest, we could use some equally-sized sample plots and count the individual birds in each plot. If we had five plots the results could be 1, 3, 12, 9 and 15, with a mean of 8 birds per plot. However, we could also have a mean of 8 birds with results of 5, 10, 7, 8 and 10. There is a smaller spread of values around the mean in the second set of data, which allows us to say that the results are more precise than the first set.

We may have a precise answer, but is it accurate? Unfortunately, in most bird censusing work we can never know the answer to this question. In the example above, perhaps some of the birds in the census plots were missed by the observer; some individuals may not have been moving or calling and were therefore cryptic (difficult to detect). If this holds for all of the plots, we may still have a precise mean estimate (all the samples are close to the mean) but it is actually a biased estimate – it is an underestimate of the real density and is not accurate. The relationship between precision and accuracy is further explained in Figure 6.

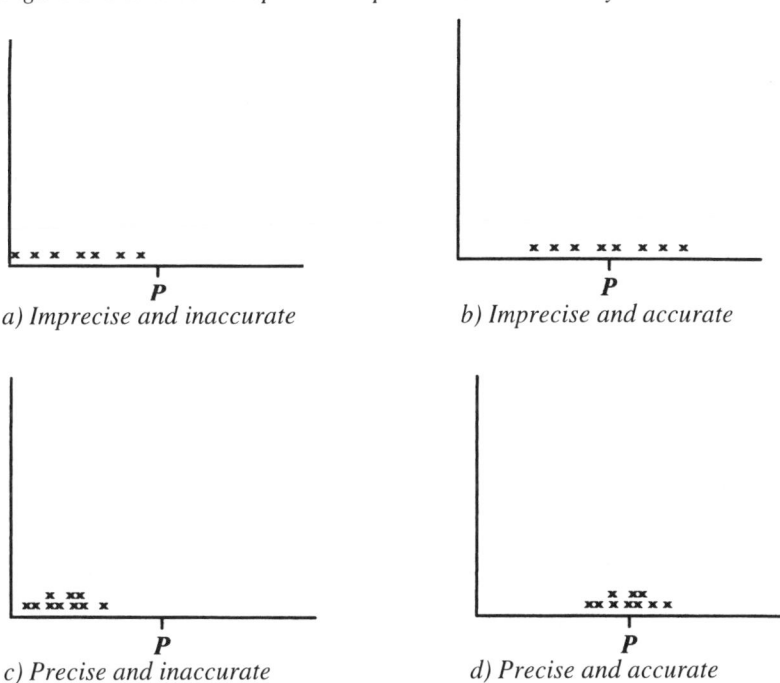

Figure 6. The relationship between precision and accuracy.

```
         |
         |
         |
         |
  x x x xx x x
         |_____
                P
  a) Imprecise and inaccurate
```

```
         |
         |
         |
         |
           x x x xx x x x
         |_____
                P
  b) Imprecise and accurate
```

```
         |
         |
         |
         |
     x xx
  xx xx xx x
         |_____
                P
  c) Precise and inaccurate
```

```
         |
         |
         |
         |
              x xx
           xx x xx x x
         |_____
                P
  d) Precise and accurate
```

The graphs shown are estimates of the density of bird species. The real density, unknown to those collecting the data, is indicated by a '**P**'. In 6a the results from the separate plots cover a wide range which does not encompass the true density - the estimates are imprecise and they are inaccurate. In 6b the estimates cover a similarly wide range but this time the true value is within that range - the estimates are imprecise but accurate. In 6c there is a narrow range of estimates which do not encompass the real value - precise but inaccurate and in 6d there is a narrow range which encompasses the real value - precise and accurate.

As we rarely know whether the answer is accurate or not, all we can do is get as precise and therefore a reliable answer as possible. If we have recognised and tried to minimise the bias in our sampling methods, we would also hope that the answer was an accurate one.

2.1.4 Relative and absolute estimates
In some cases, the accuracy of the estimate is of secondary importance or may not even be relevant. For example, if you want to know if numbers of a particular species are increasing or decreasing you could set up some census

routes, record bird contacts and use these data as a baseline to compare with data collected in exactly the same way in the future. This is a relative estimate. The actual density of birds is not important; all that matters is how one estimate relates to another. With relative estimates you may even accept some types of bias, provided the same bias is present when the census is repeated. Thus, referring to Figure 6, for studies of population change (and relative estimates in general), the data presented in 6c are as useful as those in 6d. If relative estimates do satisfy the aims of the study, it is particularly important that the methods are recorded well enough to be repeatable.

If the aims of the study require us to know the actual density of birds, then what we attempt is an absolute rather than a relative estimate, and now the elimination of bias is the most important consideration. More information on the choice between relative and absolute estimates is given in Section 2.5.

2.1.5 Measuring and increasing precision

Whether we are attempting relative or absolute population estimates, a major goal of study design is to provide as precise an estimate as possible; we therefore need some way of measuring precision. There are a number of different statistics that could be used, but perhaps the most useful thing to do is to calculate the 95% confidence limits of your estimate (the DISTANCE software discussed in Section 3 automatically calculates the 95% confidence limits of any estimate it produces). One way of defining the 95% confidence limits (although not absolutely correct in statistical terms) is to say that there is a 95% chance that the true estimate lies between the upper and lower limits. For example, a population density estimate of 250 birds per km^2 might have limits of 50 and 450, i.e. you are 95% sure that the true density is between these two limits.

Having calculated the confidence limits for an estimate, it might be apparent that the estimate is so imprecise as to be virtually useless, so how can we increase precision? One way is to increase our sample size – the more samples you take, the more precise (and more reliable) the estimate will become. Unfortunately, improvements in precision are proportional to the square root of the sample size, so to double the precision you need to increase the sample sizes fourfold.

A biased sampling procedure may also contribute significantly to imprecision of the estimates, and this also needs to be recognised and addressed. For example, if half the sampling sessions were in the morning and half in the afternoon when perhaps the birds were less active and more likely to be missed, combining data from morning and afternoon sessions would give lower density estimates but with wider confidence intervals – a less precise estimate as a result of a biased sampling procedure (more information in Section 2.2.1).

2.2 Study Design

In order to get as accurate estimates as possible, or at least to know why estimates may not be accurate, we need to identify and address any causes of bias in our sampling regime. In the unlikely event of us being able to eliminate all causes of bias, natural variation in habitats and bird distribution will still reduce the precision of the estimates. In designing a study, we therefore need to consider both the problems of bias and how we monitor and cope with natural variation.

2.2.1 Choosing the right time and conditions

Many factors will affect bird activity and behaviour, and these in turn affect your chances of actually recording the birds. Among the more important factors are time of day, the season and the weather.

Time of day

Figure 7 illustrates some of the effects of time of day on bird activity. These data on parrots and a hornbill species were collected from a vantage point overlooking a small patch of forest on the island of Sumba, Indonesia. There are morning and late evening peaks of activity with many fewer movements in the middle of the day. Many forest birds will show similar trends, and singing and calling can be even more strongly biased towards the early morning activity peak. The aim of a census may be to record as many as possible of the birds that are actually present, and usually as quickly as possible, so collecting data at the peak of bird activity can be fundamental to good study design. However, birds can be so vocal and active at dawn that it may be impossible to record all bird contacts correctly and there can be rapid changes in conspicuousness over a short time. A common study design, therefore, is to begin data collection about 30 minutes after dawn and continue to mid-morning, when bird activity declines. There may be another censusing period before dusk. As part of a pilot study (section 2.3), you could determine empirically when the best time for your own surveying would be.

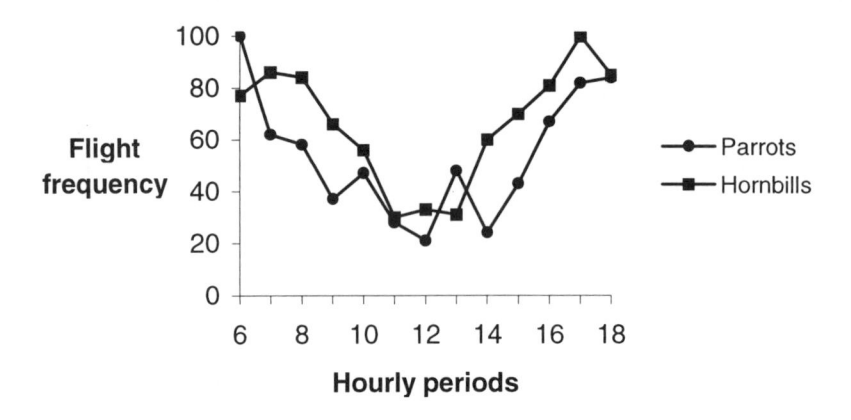

Figure 7. Changes in bird activity with time of day. Shown are the frequencies of flights of Indonesian parrots and hornbills from data collected during long watches overlooking forest patches. Flight frequencies are expressed as percentages of the maxima (0600 to 0700h for parrots and 1700 to 1800h for hornbills).

Even though censusing can be restricted to periods of higher bird activity, there is bound to be some variation of activity within the restricted period. This can be an important cause of bias. For instance, if all censusing starts at a field base and moves into the surrounding forest, all adjacent areas of forest will be censused early in the day and all distant areas later in the day. If more bird contacts are made earlier in the morning, then the adjacent forest areas will erroneously appear to have higher bird densities and diversities than the other areas. A good study design would reduce this bias by ensuring, for example, that alternate censuses were begun at the 'far end' of the routes. If census routes are being repeated, then the same route should be walked from both ends.

Season

Seasonal effects can be more difficult to cope with. Bird conspicuousness will probably change with season, and in tropical forests there may not be synchronisation of breeding cycles between or even within species. In a species which is breeding, the males may be singing and calling to defend a territory and so may be easy to record, whereas the females incubating eggs may be the opposite. There can be no hard and fast rules about whether studies are better designed to avoid or coincide with the peaks of breeding activity, as this is better determined by the aims of the study (e.g. do you want to get information on the breeding population, or perhaps the non-breeders and migrants?). The best approach to reducing bias is to collect as much information as possible on breeding activity as you progress through the study. If, at the end, you discover that all contacts were with singing males, you might be able to assume that females were incubating eggs and that your population estimate should be doubled. However, in some cases sex ratios may be unequal, and this can be an erroneous assumption. Furthermore, in a population where there are many unpaired males, there may be more song than in a healthier population where all males are paired. In some cases where you do record both males and females, but you realise they are behaving very differently, it may be appropriate to calculate densities for the sexes separately and then add the estimates together.

Weather conditions

Adverse weather conditions such as low cloud, high winds, rainfall and even very high temperatures can affect census results in three ways. Firstly, bird activity can be directly affected (usually reduced), which will affect the efficiency and reliability of your data collection. Secondly, the conditions could reduce your chances of actually seeing or hearing the birds. Thirdly, you cannot pay adequate attention to counting if you are too hot, too cold or wet. Census results can also be affected by conditions underfoot (during dry periods, fallen leaves may become very noisy to walk on), or by the noise of cicadas (whose activity is influenced, amongst other things, by temperature and humidity).

In order to reduce bias, all censusing should be carried out under a standard set of conditions, e.g. light winds and no precipitation. It is also a good idea to record weather conditions such as cloud cover, wind strength and temperature even when they do conform to your 'standard' conditions, since you might want to analyse their effects later.

2.2.2 Observer bias
Species identification

Being able to identify your target species is an obvious prerequisite for any study. Assigning contacts to the wrong species can cause under- and over-estimation of densities, as well as bias in species diversity estimates. For many forest studies, difficult identification problems are compounded by the fact that many contacts are through songs and calls. In a recent study on the Indonesian island of Sumba, the percentage of contacts for different species that were through calls rather than sightings varied from 0% to 99%, but the mean value was just over 70% (Jones, unpubl.) It may require weeks of practice to learn the calls and be able to recognise the majority of contacts.

22 Expedition Field Techniques

It will have become obvious by now (if not during the planning stages of the work) that it is impossible for an inexperienced team to visit an area of tropical forest and expect to census the entire bird fauna. An exception to this might be a project on a small island with a small number of species, but normally you will need to restrict the scope of the project. It may be better to have precise population estimates of a few key species or diversity estimates for an important guild of species than unreliable data on the whole fauna.

The magnitude of the species recognition problem is partially dependent upon how much information is already available. If your species and area are covered by a field guide and bird call tapes, you have a good basis on which to build up your knowledge. If not, you will need to collect all the available information and prepare your own version of a guide. It is often worthwhile visiting a museum and taking photographs and notes on lesser known species. Once in the field, it will be useful to involve local guides/scientists who know the birds and their calls.

Once you are at the study site, there is no substitute for good fieldcraft and bird-watching skills, taking careful notes and discussing identification problems with fellow recorders. A useful technique to employ during a pilot survey (see Section 2.3) and even during the main data collection is to plot the proportion of unknown contacts over time. Figure 8 shows such a plot for the Sumba study.

Figure 8. The decline in proportion of unidentified contacts with increasing field experience.

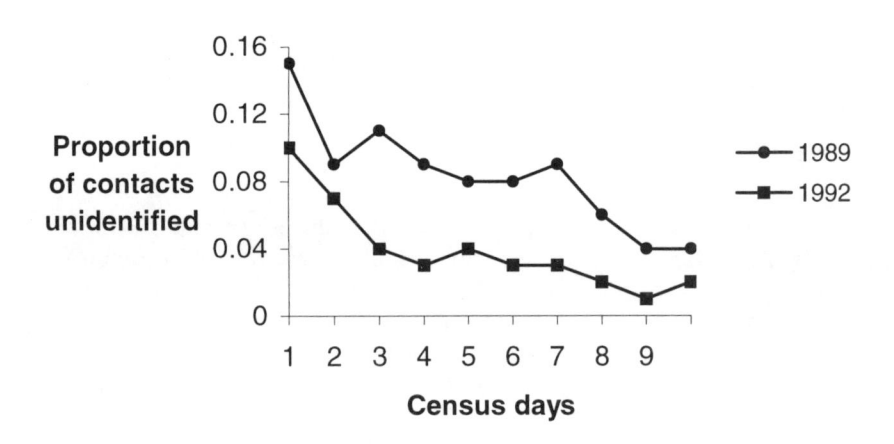

These data are taken from two visits to Sumba Island, Indonesia in 1989 and 1992. The 1989 visit was the first contact with the fauna by the survey team but the same recorders were also on the 1992 expedition. In 1989 after 4 days of experience in the field, the proportion of unidentified contacts was down to less than 0.1 (10%) and after ten days it was standing at 0.04 (4%). In 1992, because of the previous experience in 1989, fewer birds were unidentified at the beginning of the study. Many of the unidentified contacts were later identified from notes on their calls and behaviour.

Estimating distances

Some census methods require observers to estimate distances to bird contacts; these estimates can be a major cause of bias. Small random errors are acceptable, but large or systematic over- or under-estimates of distance are very serious. There are two ways to reduce these errors. The first is just to practise by selecting an object, estimating the distance to it and then checking the estimate with a tape measure. This practising can begin at home and can be finished off at the field site. When data collection is underway, it is very important to check some distances regularly to make sure there is no drift in your estimates. Everyone can improve their estimates with practice, but allow at least a week (two would be better) of regular practice and monitor how well different team members are performing (see Section 2.3). This training period is not wasted time: you need it to get unbiased density estimates and it is probably also the time when you are learning the bird fauna. The second way to reduce errors is not a substitute for the practice but makes estimation easier: under some circumstances it may be possible to use an optical range finder (not usually much use in forests) or, if you are using point counts, you could position reference markers at known distances from your census points.

It is, of course, much more difficult to estimate distances to bird calls (and most of the contacts may well be calls) and here even more practice is required. Ideally, one team member could play calls of various species from a tape recorder at measured distances, but out of sight, from the rest of the team, who then make their practice estimates.

It may seem a difficult problem to estimate distances reliably, but there are three important things to remember. The first is that, in forests, density estimates are usually based upon contacts over fairly short distances (for many flycatchers, warblers and sunbirds it is often the contacts up to 20m which are important) and these are likely to suffer from smaller errors. The second is that although it is better to estimate distances to individual bird contacts, it is perfectly acceptable to classify contacts into distance bands or even within or outside a specified distance (see Section 3). If distance estimates are particularly error-prone, it is obviously easier and also statistically better to allocate the distances in this way. The third point is that in spite of the problems, there is more information in a census with distance estimates than in one without.

Inter-observer differences
Even after a lot of training, there may still be some differences between team members in their recognition of bird species and in their estimates of distances. There will also be differences in visual and aural acuity and in powers of concentration. It is very important to be aware of, and to try and accommodate, these differences – even if you cannot eliminate the bias, it is often better for all team members to be making the same errors, rather than each member making a different error.

It is important to discover and continually monitor what the inter-observer differences might be. The best way to do this is to carry out a pilot study as part of the initial training period, and to build in further monitoring throughout the data collection period (see Section 2.3). Once the differences have been identified, a number of different options are available to deal with them. Firstly, the differences could be eliminated by further practice and/or negotiation during the training period; secondly, you could allocate different duties to different team members, e.g. the best distance estimators should be censusing birds rather than measuring tree girths; thirdly, you could organise the fieldwork so that bias is hopefully cancelled out. As an illustration of the third point, if you have two main bird recorders in the team and you know or suspect that there are differences between them, you can do two things; either make sure that all censuses are carried out separately by both recorders and the data pooled, or ensure that they carry out the censuses together and all distance estimations and species identifications are agreed between them. The wrong thing to do is to continually send one recorder to habitat A and the other to habitat B, as you will then be unsure whether any differences found were real.

2.2.3 Sample sizes and replication
Questions about sample size (numbers of contacts for a species, or number of sites sampled) relate more to coping with natural variability than observer bias. As a general rule, the more natural variability there is, the larger sample sizes you will need to get reasonably precise estimates. Birds which are clumped in distribution or usually occur in flocks will need more sampling effort as there is more natural variation in their distributions (see Section 4).

In the majority of bird conservation studies, sample sizes are too low. This is not necessarily because of a lack of effort, but because threatened species are usually rare. In most cases it is wise to collect as much data as possible; but remember that initial increases in sample size have a relatively large effect on precision while the continued increase in sample sizes has less and less effect (see Section 2.1.5). The question of when to stop collecting data may never arise if the target species are particularly rare. It may arise for commoner species and/or long periods in the field, or for diversity studies.

There are two strategies for determining your required sample sizes. The first is that some of the methods for estimating bird densities (Section 3) and examining habitat associations (Section 6) actually recommend minimum sample sizes. The second is that a pilot study and further examination of the data when they come in can be invaluable in determining how the project develops. For instance, by plotting the rate of increase of sample size for one of your key species, you can estimate how long it would take to get the minimum recommended sample size. If this time is beyond the study period, you could either accept that you will never get the optimum sample and concentrate on the study's other aims, or redesign your fieldwork to try and get bigger samples, e.g. perhaps you could spread your sampling over a wider area. The sample sizes you aim for and ultimately accept as being adequate will depend upon the aims of your project. Sometimes, an order of magnitude for a density or diversity estimate will satisfy your aims so you may only need a small sample and a fairly imprecise estimate before moving on to another area. As mentioned in Section 1, as few as ten contacts with a species can be enough to make some sort of estimate of its abundance. It may not be a particularly precise estimate, but it may be adequate to fulfill your particular aims.

Replication of your sampling is a way of increasing the reliability and general applicability of your results. There are two sorts of replication: one involves resampling the same sites; the other replicates the whole study at another site. Resampling the same sites would involve repeating the censuses you have carried out at particular point counts or line transects. These replicates have to be treated as such, rather than as independent samples. This type of replication can be a good way of getting more precise information in restricted areas and increasing sample sizes for density estimation (see Section 3). It can also be organised in such a way as to allow you to check for bias and consistency. If you have time, it is always a good idea to do at least some replicates.

Replication in its other sense can be illustrated by the following example. One of the aims of your project might be to compare the species diversities of three forest habitat types in a protected area. You design and carry out the study in the appropriate way, and you may even be employing the type of sample replication outlined above. After analysing the data, you may be able to say that the avifauna of habitat x is more diverse than that of habitat y or z. This might well satisfy your particular aims relating to the management of that particular area. If your aims are actually more general (e.g. is habitat x always more diverse than y or z?), then it would be unwise to base your conclusions on the results from just one site. Replicating the whole study at other places where habitats x, y and z occur together would confirm whether there are consistent differences between the habitat diversities, or whether any differences are site-dependent. The decision to replicate individual samples or the whole study in different areas depends upon your particular aims, but such replication is a powerful tool and one which is too often neglected in conservation studies.

2.2.4 Positioning your sampling effort
Habitat heterogeneity
In the discussion of sample sizes above, it has almost been assumed that it is the total sample size which is important. What is more important is the sample size within each sampling unit or habitat. Combining data from different habitat types will provide bigger sample sizes, but if birds are not distributed similarly between them, you will get biased estimates and problems with precision and accuracy.

The first step is to establish how many broad habitat types you have in your study area. You can never do this in an entirely satisfactory way, as you will not know the important habitat divisions and gradients as far as the birds are concerned. Nevertheless, even the broadest distinctions will reduce bias and hopefully increase the precision of your final estimates. The main habitat types can be identified from standard maps, aerial photos, satellite maps, pilot surveys and local knowledge. Since the main habitat types are not necessarily the smallest sampling units, it is probably wise to treat areas with the same habitat, but which are geographically distinct, as separate sampling units. It is always possible to examine the data from the two areas at a later date, and if there are no obvious differences, combine them for further analysis.

Once the smallest sampling units have been identified, you can then sample adequately within each unit. What constitutes an adequate sample depends on the aims of the project, the degree of natural variability present and the methods you are employing, but a rough guide would be to aim for about 50 point counts or 10km of line transect within each sampling unit.

If a main aim of the project is to establish habitat associations and preferred sites, it is essential to sample over as complete a range of habitats (and usually altitudes) as possible, and to include areas where the target species may be rare or even absent.

Positioning of sampling sites

If your sampling sites are specific points, then the best way of positioning those sampling sites within the sampling units is probably through a stratified random technique. This involves dividing up the study site with a grid, either on a map or actually on the ground with markers, and then using random coordinates to position the sampling site within each grid square (see Figure 9). Unfortunately, in many cases there will neither be an adequate map nor the time or resources to set up a grid on the ground (although the latter is undoubtedly the best option for longer-term studies). It may be possible to position sampling sites randomly in other ways, but for safety reasons this may not be a good idea – it is easy to get lost if observers are leaving paths and searching for randomly selected points.

If you are going to sample by continuously walking along transects, you could start the walks from points which are determined randomly or systematically, and the direction you walk could be random or systematically organised. If you adopted either of these approaches in a forest habitat it would take a lot of time, effort and habitat destruction to cut the trails you needed. Although this might be the best approach for a longer term study, in practice you may have to compromise and follow existing paths, stream beds, etc. You will obviously save time doing this, but you are almost certain to get biased data. The clearance and continued use of paths and the presence of a stream or river is bound to have an impact on the surrounding vegetation. The initial siting of a path is also not likely to be random with respect to topography and vegetation.

By just collecting data along existing paths, bird and plant communities characteristic of forest edges will be over-represented in the data collected. One compromise would be to census along existing paths, but to make short forays (for line or point counts) away from the paths at randomly determined intervals. You could then compare the data on and off paths and assess how biased your total data set might be.

If you are adopting a non-random approach to positioning sampling sites you must be fully aware of the potential bias in the results. A useful approach is to identify the highest environmental gradients within your study area (e.g. low to high altitude or open to closed canopy) and deliberately sample across those gradients. Observing and analysing the trends along the gradients will help you understand the biases within your whole data set.

Figure 9. Positioning point count sites or the beginning of transects.

a) Along paths or rivers *b) Random*

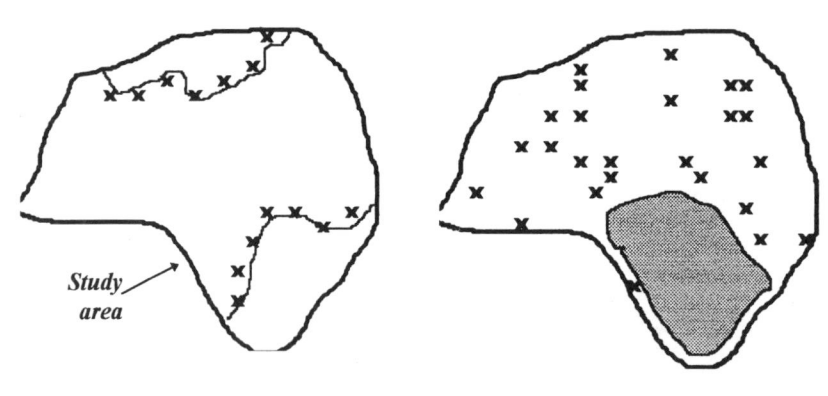

c) Stratified–random method

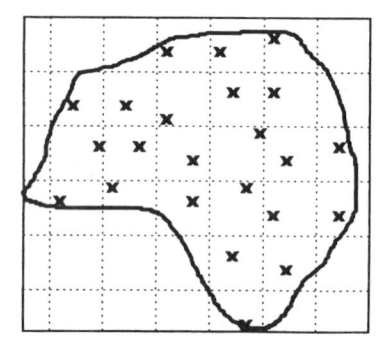

In (a) point count sites have been positioned along paths or rivers. This has the advantage of easy access and relocation but only parts of the study area have been sampled and edge habitats are likely to be disproportionately sampled. In (b) the point count sites have been chosen randomly which has the advantage that the study area will be more evenly sampled. However, it is possible that a completely random choice might leave some areas undersampled (indicated by the shading in (b)). The best method is to use a stratified-random method to place the point count sites. The first step is to superimpose a grid onto the study area. This could be done on a large-scale map and interpreted on the ground with a Global Positioning System, or actually marked on the ground itself. For the latter you only need to mark one corner and decide upon the compass orientation of the grid. The distance

between the grid lines will depend upon the distance over which birds can be detected but you will probably need at least 500m. Within each grid square one or more positions can be selected using random numbers. These positions could be used as point count sites or as the starting points for line transects (which can then all proceed in the same direction or a randomly determined compass direction). If a particular grid square encompasses ground outside of the study area, keep taking pairs of random numbers until the site indicated is within the study area. Similarly, if sites chosen in adjacent squares are very close (and the same birds could be recorded from two points) only 'accept' the random numbers if they place the sites more than a set distance apart. This distance depends on the distance at which birds are detected but in forest this may be 150 to 250m.

2.3 Pilot surveys and training

It is obvious from the previous section that, in order to get reliable estimates, you need a carefully designed study and well trained personnel. Many aspects of study design and part of the training can be accomplished before beginning the fieldwork, but much still needs to be done when you arrive at the field site. You may need to allow at least two weeks for further training and a pilot survey to refine the study design (and even the aims of the project). At this stage, liasing with or actually employing local experts becomes particularly important.

The first stage of the pilot study is to familiarise yourself with the bird fauna and the habitats:

- obtain as much local knowledge as possible on the distribution of key species and habitat types;
- make detailed notes on sightings and calls, comparing them to taped calls if available;
- compare and discuss identification problems between observers;
- start a daily log of bird records;
- begin to practise distance estimates (if these are appropriate for your study);
- start tree species identification or classification (if appropriate);
- start to plot habitat boundaries, likely census routes and other important features on available maps, or begin to construct your own maps;
- decide upon the smallest sampling/habitat units.

For the next stage it is a good idea to set up one or two census routes (either for line transects or point counts) and select one person to organise and monitor the performance of the rest of the team. If it was not obvious before the fieldwork, it should by now be obvious who the best bird identifiers are, so agree a division of tasks among the group:

- get the bird recorders to repeat the same census routes – are they recording similar numbers of contacts per species? If not, get them to census together to sort out any problems;
- monitor the proportion of unidentified bird contacts over time – is this proportion declining fast enough? If not, put more work into species identification problems or redefine the aims;
- compare the performance of different team members in estimating the distance to known objects – who is the most accurate, who are over- or under-estimating? Is more practice required? Should only certain individuals be 'allowed' to make the estimates?
- similarly, compare abilities to estimate tree heights or identify important habitat features;
- identify and agree reference points for particular habitat variables, such as canopy cover;
- repeat your practice census routes at different times of day, or monitor activity for long periods from vantage points, and decide upon the times of day when data collection will take place;
- for the key species, monitor the initial census results and predict how long it will take to reach the required sample sizes;
- for each key species, calculate the mean and standard deviation of the number of contacts per point count or section of line transects. For the species with the largest standard deviations (the less precise ones), plan to get larger sample sizes (likely for flocking species or those with uneven distributions within the same habitat);
- calculate some species discovery curves (see Section 5) to estimate how long you need to stay at sampling stations and within sampling units.

At the end of the pilot project, it should now be possible to decide the following:

- the positioning of the sampling sites (point counts or line transects);
- the make-up of sampling teams and the division of responsibilities within the whole group;
- whether any of the project aims have to be refined (e.g. you may have discovered that you do not have enough time to get reasonable data from six sites, so perhaps plan to visit the four most important ones).

It is also a good idea at this stage to produce a standard sheet for recording the data. A generalised data sheet should have been designed and copied before getting to the field site but you may need to refine it now. Do not allow team members to use their own notebooks for collecting census data. Without the appropriate headings (and reminders), you always lose information.

Once serious data collection has started, the procedures adopted during the training period should not be completely abandoned. You should still monitor things like the proportion of unknown contacts, distance estimating abilities and how sample sizes and levels of precision are developing. Monitoring sample sizes is an important factor in deciding if and when to move to a new field site.

2.4 Safety, team size and logistics

Having designed the study and trained your team appropriately, you may still have to make concessions for health, safety and logistical reasons.

2.4.1 Health and safety

For safety reasons (and for the division of data collection tasks – see 'What size team?' below) data collection teams should always comprise of at least two, preferably three, people. Potentially dangerous areas should be avoided, however interesting they look – you cannot concentrate on looking for birds if you are watching every step or hanging onto a steep slope. Everyone needs time to rest, so do not plan for everyone to collect data every day; you have to be fit and alert during data collection. Be prepared for the fact that most team members may be ill for at least some of the time, so build plenty of 'slack' into your study design.

2.4.2 Logistics

Generally, there is a trade off between the amount of time spent travelling and collecting data. There may be lots of potential places to visit, but good data from a restricted area may be better than poor data from a wide area. What you choose to do depends upon the aims of the study: looking for a rare and little-known species may require you to cover a lot of ground; getting precise density or diversity estimates, or detailed habitat association data, usually means more time spent in fewer places. If you can afford to do so, hire people to do as much of the ancillary work as possible. If much of your equipment is carried for you, and you have cooks to buy and prepare food, you will be able to put more of your effort into the data collection. Such local collaboration will pay other dividends (see Section 7).

2.4.3 What size team?

There are two aspects to this question. The first concerns the size of the data collection teams who are actually carrying out the censusing, and the second concerns the size of the whole project team. The advantages of having a large group are that you can collect more data; you are covered for illness; you can involve more local collaborators in the project; it is probably cheaper per individual and, having arranged visas and permits and transport, why not take as many people as you can? The main disadvantages are that it is logistically more complicated to move and feed large groups, you may have a larger negative effect on the local environment and you may need particular experience and skills for organising a big group.

For data collection itself, teams of three (two observers and a data recorder) are probably the best for the following reasons:

- the two observers can concentrate solely on identification and distance estimation;
- identification and distances can be compared between the two, so individual differences can be evened out;
- the recorder can check that all the information for each contact has been provided;
- the recorder can be another check on the distance estimates and can make sure the observers are concentrating!
- there are safety advantages of having three – following an accident one person can go for help whilst another can provide first aid.

The disadvantages of having three people per team, rather than one or two, is that larger groups make more disturbance and it is obviously less efficient than having teams of two – potentially you will be collecting fewer data. Obtaining complete and less biased data in a safe way is probably the more important consideration, but if you do have teams of two, make sure that an experienced person is combining the surveying and recording jobs. Whatever size you do start off with, you should maintain it for the whole study period.

Although three is arguably the ideal size team for data collection, it is too low for the group as whole. Having only three would leave little time for collecting habitat and other useful data, with no scope for domestic logistics and health problems. Four or five would be a more realistic minimum team size.

2.5 Which methods to use?

2.5.1 Introduction
The methods adopted will depend upon the aims of the project. As a general rule, adopt the simplest methods which satisfy those aims. The more complex methods will usually be more demanding in time and in the statistical criteria

you have to satisfy. It is better to get reliable data using a simple method than unreliable data from a complex one, even if the latter (potentially at least) could provide more information. Another reason for adopting simple methods is that these are more likely to be repeatable. If you hope that others will repeat your work in the future, perhaps as part of a long-term monitoring programme, do not assume that they will have the same level of training or put in the same amount of effort as you. Local conservation workers often do not have the time to repeat complicated surveys, so keep the aims and methods as simple as possible.

Different methods apply if you are interested in bird diversities, species densities or habitat associations. For bird diversity and bird/habitat methods, go straight to Sections 5 and 6 respectively. For bird densities, a number of decisions about study design have to be made before moving on to Section 3.

2.5.2 Bird densities

The basic decision to be made first is whether you want relative or absolute density estimates (the difference between them was outlined in Section 2.1.4).

If you want to know if numbers of a particular species are increasing or decreasing, or if you want to compare the birds in two areas of similar habitat, then relative estimates may satisfy your aims. In generating these relative estimates, you would need to standardise your methods and get as precise estimates as possible. Potential causes of bias should be identified, but you may decide to ignore some of them as long as the same bias is present in all the areas you might be comparing.

Relative estimates do not allow you to make comparisons between species. This is because different species have different levels of conspicuousness (or call output). The same problem exists when comparing the same species between different habitats – the species may be more conspicuous and therefore appear to be commoner in one habitat than another, when in fact the only real difference is that it is easier to record in one of the habitats. It is also in the nature of relative estimates that you cannot derive population density or sizes from them.

Distance sampling (Section 3) involves estimating distances to bird contacts and, theoretically at least, provides absolute density estimates from which you can derive population sizes for particular areas. In practice, because of all the sources of bias which can affect accuracy, we can not easily know how close our 'absolute' estimate is to the real figure. However, a very important extra reason for using a distance sampling method is that it allows for different levels of conspicuousness between species, and between different habitats occupied by the same species. At least you will then be able to say that one species is probably more common than another, or that a species is commoner in habitat x than it is in habitat y.

2.5.3 Point counts and line transects

Throughout this section two different methods of censusing have been mentioned – point counts and line transects. The former involves walking to, and usually marking, a particular spot, and then recording all bird contacts for a pre-determined period (often 5 to 10 minutes) before moving on to the next point. Line transects involve the observer continually walking and recording all contacts either side of the track walked. The precise details – including how long you should stay at a point, how far the points should be apart and exactly how you collect the data on line transects, etc., are discussed in Section 3. Whether you adopt point counts or line transects depends upon a number of factors. The advantages of each method and, implicitly, the disadvantages of the other method, are given below.

Point Counts:
- concentrate fully on the birds and habitats without having to watch where you walk;
- more time available to identify contacts;
- more likely to detect the cryptic and skulking species;
- easy to relate bird occurrence to habitat features.

Line transects:
- cover ground more quickly and record more birds;
- less chance of double recording the same bird;
- good for more mobile, more conspicuous species and those which 'flush' easily;
- errors in distance estimation are less serious than for point counts (see Section 3 for explanation).

If you are targeting a few species which are relatively easy to identify but which may be mobile and occur at low densities (usually larger species, such as parrots), line transects are undoubtedly better. If you are censusing a larger element of the bird fauna and especially if the species are small, flocking and difficult to identify, then point counts are better. There are many studies, of course, for which the choice is not straightforward and perhaps neither method is ideal. Section 4 gives more information on what to do with these difficult species.

Section 3
ESTIMATING BIRD DENSITIES USING DISTANCE SAMPLING

Huw Lloyd, Alexis Cahill, Martin Jones and Stuart Marsden

3.1 Introduction

In Section 2, the distinction was made between censuses that provide a relative measure of bird abundance (e.g. numbers encountered per hour or per km) and those that produce an estimate of bird density (number of birds per unit of area). Of course, estimates of actual bird density are only needed if the aim of the study is to produce them, to use density data to calculate total population estimates, or to relate your figures to those of past surveys where density estimates were calculated. However, these are usually very good reasons for using distance sampling and it should be remembered that it often takes little longer to collect 'distance data' than it does to collect data using other methods. What does take time is the planning and practice needed to collect reliable and meaningful data.

The general way of producing density estimates is through 'distance sampling' (other methods are outlined in Section 4) and this can take place on point counts or line transects. The crucial part of the method is that an estimate is made of the distance from the bird contact to the centre of the point count site or to the line which a transect walk is following. These distance estimates are used to calculate bird densities and, of particular importance, they take account of the fact that some birds are detectable over much greater distances than others, and that a species may be more easily detected in one habitat than another. Thus, even if calculating total population sizes is not the main aim of the project, collecting the distance data will allow you to make direct comparisons between species and between the same species in different habitats. These are comparisons which may not be possible with encounter rate or other relative density estimation methods.

There are four basic assumptions of distance sampling that should be adhered to if an unbiased density estimate is to be obtained:

- transects or points are representatively placed with respect to bird density;
- objects (birds) directly on the line or at each point are always detected;
- objects are detected at their initial location prior to natural movement or movement in response to the observer's presence;

- distances should be accurately measured (or at least estimated with small and random error).

This section aims to provide a basic understanding of distance sampling methods using both line transects and point counts, and also to show how a study using distance sampling should be designed to meet the four critical assumptions listed above. Once the relevant data have been collected, it is possible to calculate approximate density estimates with a calculator but recently a computer program has become available which produces the estimates in a more sophisticated way. This program is called 'DISTANCE' and is freely available. The program has a companion book called Distance Sampling (Buckland *et al.* 1993). In this section we will explain how to use the software to analyse your data in what we suggest is the most appropriate way. Following this introduction, and certainly if you intend to publish your findings in a scientific journal, we strongly recommend that you get hold of the distance sampling book and the manual (Laake *et al.* 1994) which accompanies the DISTANCE software.

First, we introduce the two main methods of collecting distance sampling data, using line transects and point counts and discuss some of the options available to minimise problems which can arise during data collection. Next, we discuss some of the problems of calculating density estimates from the data you have collected. We introduce the workings of the DISTANCE program itself and include some example syntax and output from the distance program (Section 9).

3.2 Distance sampling using line transects

3.2.1 Introduction
The choice between line transects and point counts has already been considered in Section 2. To summarise, line transects may be better for lower density, more mobile species in fairly even habitats. Point counts are better for skulking species or for censusing larger numbers of species and for work in fine-grained habitats.

3.2.2 Positioning of transects
It is best to site the start of the transects randomly or through a stratified random technique (see Section 2). This is one of the four basic assumptions of distance sampling: line transects that are randomly placed with respect to the distribution of birds are more likely to produce unbiased density estimates which can be extrapolated to other areas of the same habitat type. If the location of line transects is chosen subjectively, or for the observers' convenience (e.g. along trails or in an area which appears to contain high

numbers of birds), the sample obtained is only strictly representative of the area surveyed. Usually, for logistical and safety reasons, transects are not randomly situated and it is important to be aware of how this may bias the results. Walking transects along large rivers or wide trails/roads may be a particular problem as the vegetation to each side of the transect route may be highly uncharacteristic of the forest as a whole (see Section 2).

Sometimes transects are laid out in grids which are oriented to a contour or obvious feature in the landscape, such as a road or a river. Using such grids may not provide a random sample but it may be fairly easy to identify and test for causes of bias (e.g. transects near to rivers can be compared to those further away). They may also be very useful for long term studies where population changes are monitored at one site.

3.2.3 How many transects and how long should they be?
The total length of line transect in a study depends upon how long it takes to get an adequate sample size for the target species and how many habitats are to be sampled. At the end of your pilot study, you should be able to predict how long it will take you to collect enough data and therefore how many kilometres of transect will have to be walked. The longest transect walked in any one day is not likely to be more than 10km. This is because censusing is often restricted to periods of high bird activity, and the quality of the data collected will decline as the observers begin to tire. If you need precise estimates in well defined areas or habitats, it might be better to do many short transects of, perhaps, around 4km. It then becomes easier to avoid some of the bias related to time of day.

Each transect can be partitioned into distance intervals along its length. For example, markers every 50m along a transect can help the observers to follow the correct track and also allow habitat information to be collected for specific sections of the transect. The habitat data can then be related to the occurrence of bird species at particular sections of the transect (see Section 6).

3.2.4 Collection of data
Once transects have been selected, data collection can begin. The design of the study and the methods employed should now be relatively clear. A poorly designed study will not only lead to unreliable results but also problems with using the DISTANCE computer program.

On each transect, the observers walk at a fairly constant speed, looking either side of the line walked, and estimate the perpendicular distance from the line to each bird contact. There are two ways of estimating the distance: 1) you can make a direct estimate of the distance between the bird and the line, or 2) you can estimate the distance between the observer and the bird, and the angle of the sighting away from the line. These methods are illustrated in Figure 10.

Figure 10. Distance estimation/measurement along transects. Either the perpendicular distance (d_1) from transect line to object is estimated or measured, or d_1 is calculated using d_2 and the sighting angle θ ($d_1 = d_2 \cdot \sin(\theta)$).

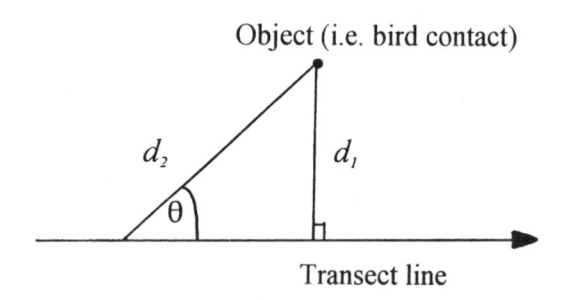

Object (i.e. bird contact)

d_2 d_1

θ

Transect line

A critical assumption of the method is that all birds at distance 0m are detected. This can be a problem if there is a dense and high forest canopy and under these conditions perhaps one of the observers should concentrate solely on the canopy. It is also important that the observer does not flush birds from or onto the line transect ahead. Although this is an important assumption of distance sampling, it can sometimes be a difficult problem to overcome in the field. More commonly, birds will be flushed away from you, so keep an eye on the line of travel ahead of you and try to record the positions from which the birds are flushed.

Distance sampling methods aim to produce a 'snap-shot' of all the birds recordable from the transect line. This creates a problem for the recording of flying birds (i.e. those not seen to leave the immediate area of the line transect), as it is impossible to know if those birds are normally part of the population of that area. Although it is worthwhile recording these observations, they should not be used in the calculations as they would produce overestimates. Leaving them out might actually cause an underestimate but the error will almost always be much smaller. Remember that if birds are seen to take to the air, then these birds should be included in the count and an estimate of distance is made from the take-off point perpendicular to the line transect.

A distance estimate and a count of the number of birds in each contact are all that are required to calculate density but it is also useful to record the following:

- the sex of the individual birds (if possible);
- the type of contact, e.g. was it a visual sighting or was the bird singing, calling, or flying?
- the time of day of each contact;
- the height of the bird e.g. ground, low, mid-strata or canopy.

This information can often throw light on the biology of target species, and is also useful when it comes to analysing and interpreting the results. For example, if for one species all the males are singing contacts and the females sight-only contacts, it is probably a good idea to carry out separate density estimates for each sex.

Example data collection forms for the variable distance line transect (VDLT) method are shown in Figure 11.

Figure 11. Example data collection forms – transect methods.

DATE: 23/8/96 WEATHER: HAZY START: 0650 FINISH: 1015
OBS: SM + MJ.   SUN.

Transect Number	Habitat type	Species	Group size	Perpendicular distance
1	PRIMARY	FB	1	15
1	"	CJ	3	9

Tran No.	Habitat	Species	Group size	Angle	Distance	Height
4	AGRIC	BHB	2	70	11	7
4	"	GIANT PITTA	1	55	4	0

3.2.5 Double counting

Counting the same bird twice can have important consequences, but as long as the detection is not within the same sampling effort (i.e. along the same line transect), then double counting will have a minimal effect on density estimates. Also, there is no problem if a particular bird is stationary and is detected from two different line transects. It only becomes problematic if that bird moves from one line transect to another after it has been initially recorded. It is obviously important to keep a mental note of bird movements and try to avoid double counting, particularly within the same line transect.

3.2.6 Variable distances and distance bands

The method assumes that the distances to bird contacts are accurately measured or that they are estimated with only small and random errors. It is particularly important for contacts near the line to be estimated correctly. Large errors or consistent over- or under-estimates will seriously bias the estimates produced by distance sampling. The importance of adequate training in distance estimation has already been emphasised in Section 2.

Estimating exact distances to individual bird contacts perpendicular to the line transect is, statistically, the most robust approach for distance sampling along transects and it is the one we would recommend. However, estimating exact distances to bird contacts can be difficult – particularly for bird calls in dense habitats. An alternative method is to employ fixed-width transects, where birds are recorded within just two or three belts of fixed distance either side of the transect. Using the Fixed-width Line Transect method (FWLT), all birds are counted along the guidelines of the normal line transect method, but each detection is attributed to a distance belt. With this method, errors in distance estimation will only have an effect if the contact is assigned to the wrong band, whereas with the VDLT method all errors have an effect. Another potential source of bias with the VDLT method is that when trying to estimate exact distances there is often a tendency to round off estimates to the nearest five or ten metres. A quick examination of data collected in a pilot study will show if this is happening. If it is, you can either try to be more exact in your estimations or adopt the FWLT method.

Two distance belts is the minimum required for density estimation but it is better to have more and it is usual to vary the widths so the bands closer to the line are narrower. You could, for example, have 5, 10, 15, 20, 30, 40, 60, 100, 200+ metre limits. In dense habitats where most of the bird contacts will be close to you, it is better to have narrower bands. The more bands you have, the better for the analysis, but the greater the problem of assigning the bird contacts correctly. If you have only two bands, the inner band should include at least 50% of all contacts in order for you to get reasonable estimates.

3.2.7 Distance estimations to groups

Sometimes it may not be possible to estimate distances to all individual birds. Populations of many species naturally aggregate into flocks or clusters. If this is the case, and you are using the VDLT method, distance estimations should be made to the geometric centre of the cluster. If using the FWLT method, all members of a cluster are assigned to the distance band which encompasses the centre of the cluster. If a species is known to always occur in flocks at the time of the census there is an extra problem: a flock may be contacted through calls but not seen and the observer may not know how many birds are present. In these cases, it is normal to substitute the mean flock size for the visually recorded flocks. See Section 4 for more information on problems associated with flocking.

3.2.8 Sample sizes

Sample sizes for line transect distance sampling data have to be quite large. Small sample sizes contain little information about density and their precision is poor, regardless of the analytical method used. An ideal minimum should be approximately 60–80 records but an estimate (albeit an imprecise one) can be calculated with fewer observations. If birds are clustered, the sample sizes would have to be even larger.

3.3 Distance sampling using point counts

3.3.1 Introduction

The difference between line transects and point counts is that, for the latter, an observer stands still in one particular location (a census station) recording all the birds seen and heard during a fixed count period. Point counts are often preferred to line transects when surveying less mobile bird species, and in more fine-grained habitats. This is because a randomly placed transect route might only pass through two or three habitat types in an area which has many more. Census stations which are randomly or systematically allocated in the same area are more likely to sample a wider range of the habitats present. Also, if detailed habitat associations of bird species are an objective of the study, habitat data can be recorded around each census station and can be easily associated with the presence/absence of individual bird species (see Section 6).

Point counts are also preferred to line transects in closed forest habitats with high canopies, particularly rainforests. This is because by standing in one location over a fixed period of time, an observer has a better chance of detecting birds than if he or she is just walking through the area. Laying out point counts systematically along transects also overcomes the problem of trying to walk and survey birds in very difficult and uneven terrain: at the census stations, time is spent trying to find birds rather than watching the path of travel (although the time walking between census stations is then 'lost').

3.3.2 Location of census stations
As with the siting of transect routes, point count sites should be positioned randomly within your sampling units or habitat types. In order to get adequate coverage in each unit, you could adopt a stratified random technique as outlined in section 2.2.3. The problems with a random placement of sites are logistics and safety. In some areas and habitats, it might be difficult and time consuming to get to all of the sites and there is a danger of becoming lost. If point count sites are positioned along transect routes, time is used more efficiently, but you must be aware of the bias that might be caused by sampling sites in a particular order and along habitat edges (see section 2.2.3). A practical way to position point count sites is to set them out along transect routes which follow trails or streams but to place each site at some distance perpendicular to the transect route itself. Jones *et al.* (1995) used this type of procedure: every other census station was placed 50m to alternate sides of the transect route.

Another important consideration is the spacing of census stations. If census stations are too close together, birds can be recorded from one station and then have a good chance of flying the short distance to the next census station. If stations are too far apart, then too much time is wasted walking between them. As an approximate compromise, the minimum distance between stations in dense forests should be 200 to 250m. If the study focuses upon small, fairly sedentary and inconspicuous birds, the distance can be smaller (e.g. 150m). For larger, more conspicuous and more mobile species and particularly for studies in open habitats, the distances should be greater – 350 to 400m is not unusual. A final decision on the spacing of stations should be made at the end of the pilot study, once you have experience of the distances over which individual birds can be recorded.

Spacing out point counts is easy if you are siting them along transect routes; it is more difficult if you are aiming for a random distribution as, by chance, two stations could be sited very close to each other. In these circumstances it is best to constrain the randomisation process, such that stations are sited randomly but, if any two are within the minimum set distance, a new set of coordinates is allocated for one of the stations. This is repeated until all stations are more than the minimum distance apart.

3.3.3 How many census stations?

This will depend upon the sample sizes required for each target species, and can be predicted from a pilot survey (section 2.3). You will need a minimum of about 50 point counts to sample the commoner species within a sampling unit (a single habitat type at one site) and to describe the bird community of the habitat. For rare species, very many point counts are needed to amass enough bird records to produce precise estimates, simply because the species is not recorded at the great majority of the point counts. The precision of the density estimates can be increased by repeated data collection at census stations (see section 2.2.3), but this is obviously at the expense of the area that could be covered during a survey.

3.3.4 How long should the count period be?

This is a difficult problem. The ideal scenario is to have an instant 'picture' of all of the birds at or near the station. In reality it takes time to detect and take details of all of the birds at the station. Even large and colourful birds may only be detected if they move or call, while cryptic birds and those high up in the canopy, may take even longer to be detected. It is a critical assumption of distance sampling that all birds at distance 0m should be detected, and it helps if there is a near-certainty of detection for some distance from the census station. Staying longer at a station should increase the chance of detecting birds but we then come up against another important assumption of the method, namely that individual birds are not counted twice (at least during the same point count). The longer the count period, the greater the chance that a bird would be counted twice or, just as importantly, a bird could move undetected into the sampling area from outside. Both of these circumstances would lead to an overestimate of the number of birds using an area at an instant in time.

Most studies use a count period of between five and ten minutes; the more mobile and conspicuous your target species, the shorter time you should use. For multi-species surveys, where different periods would be appropriate for particular groups of species, you could adopt a longer period (e.g. ten minutes), but record the time each bird contact is made. This enables you to use, for example, the first five or six minutes for the more mobile species (for which double counting could be a problem), and the whole ten minutes for the more cryptic and sedentary species. Using more than ten minutes will not usually be necessary. More information on counting periods appropriate for different types of species is given in Section 4.

3.3.5 Data collection

The variables recorded are almost identical to those recorded for line transects (see Section 3.2.4). Before the observers begin to record birds at a census station, it is a good idea to wait a few minutes so the resident birds can settle down after the disturbance produced by your arrival. Once this period is over, the observers stand still at each station, record their start time and then estimate distances to all bird contacts. Remember that you should estimate the distance of each contact to a designated point and not to the observers, who may not be standing on that exact point. It is often useful to record the exact time of each contact or to assign them to a one or two minute block of time. Information on sex, type of contact, height of contact in the foliage and group size can be recorded in the same way as with line transects. Birds that fly away from the immediate area are recorded and a distance estimate made to their point of departure. This also goes for birds flushed as you arrive at the station. Birds that fly into the area and land, or fly over the area, can be noted but should be excluded from the data analysis. Double counting of birds has the same consequences as that stated for line transects in section 3.2.5.

3.3.6 Variable distances and distance bands

As with line transects, it is important that distances should be estimated accurately or with small and random error. Density estimates generated from point counts are even more susceptible to bias arising from inaccurate distance estimations than those calculated from line transect data. This is because the total area surveyed using point counts is proportional to the square of the distance from the observer. With line transects, the area surveyed is only linearly proportional to distance from the observer. This places even greater importance on the need for accurate distance estimation and adequate training before the real survey begins.

The best point count distance sampling method involves estimating the actual distance to each bird contact, and this is often called the Variable Circular Plot (VCP) method. Contacts can also be assigned to distance bands in the same way as outlined for line transects in section 3.2.6, and the procedure adopted for recording distances to flocks of birds is the same as that outlined in Section 3.2.7. An example data collection form for the VCP method is shown in Figure 12.

Bird Surveys 45

Station No.	Habitat	Start time	Time period	Species	Group size	Distance
28	1°	0840	1 (0→2 mns)	T.h	1	12
28	1°	"	2	E.r	2?	35

Figure 12. Example data collection form – variable circular plot method. In this example, the total count period (10 minutes) has been divided into five two-minute periods.

3.3.7 Sample sizes
Sample sizes for point count data must be larger than corresponding ones for line transect data to get the same degree of precision. Ideally you should aim to accumulate 80–100 contacts for each species in each sampling unit. It is possible to calculate estimates from much smaller samples, but these will be less precise. Again, it is important to define the level of precision you need from your density estimates before you start the survey. This will help you to ensure that your estimates are precise enough to detect density differences with confidence, but to avoid spending time in collecting extra distance data, when you could be collecting other data.

3.4 Examining the data

3.4.1 Bird detectability and the detection curve
Whether one walks around a forest or stands at particular locations, an assumption of distance sampling is that all birds at a distance of 0m are recorded. Usually all birds at some distance away are also recorded, but as distance increases there is an increasing likelihood that birds will be missed. A typical 'fall-off' in detection with distance is shown in Figures 13a–b (for an African hornbill). These are data from a line transect where the numbers of bird contacts in each distance band have been totalled (distances either side of the line have been combined). Note that the two histograms are the same shape (each distance band has the same area). The birds could have been assigned to these bands during the fieldwork, or distances to each bird contact could have been recorded exactly, but allocated to the distance bands afterwards (this is done for the sake of the demonstration; they are kept separate for the actual analysis). As we would expect, more contacts are made closer to the observers and the numbers tail off with distance. Figure 13b is equivalent to the 'distance function' or 'detection curve' for species x, and describes its detectability for that particular habitat.

Figure 13. Bird detection curves from transect and point count methods. Histograms (a) and (c) show the actual numbers of birds recorded in each distance band, while (b) and (d) show the density of birds recorded per distance band (i.e. number of birds divided by the area within the distance band and expressed as individuals per km^2). Note that histograms (c) and (d) are different shapes because the area within distance bands increases exponentially with increasing distance from the recorder.

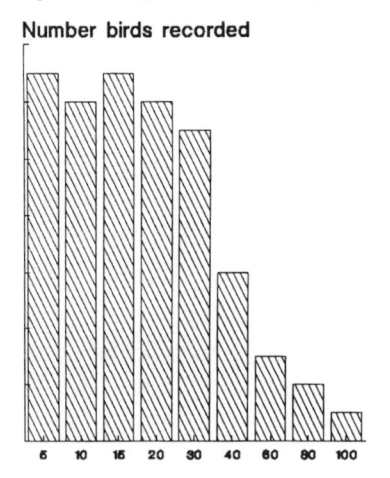

a) Transects – birds recorded

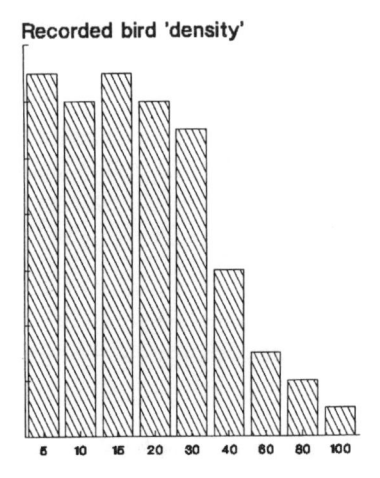

b) Transects – detection curve

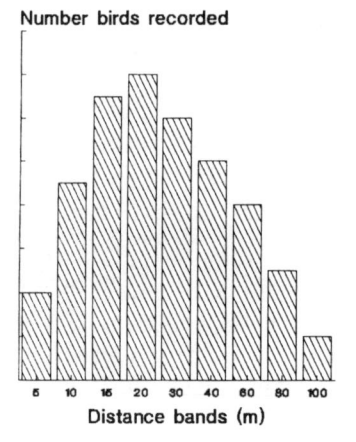

c) Point count – birds recorded

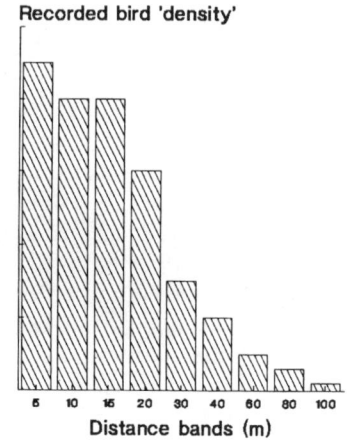

d) Point count – detection curve

Detection curves for point count data are similar, but there is the added complication that the area within each distance band is different (Figure 13c–d). The area encompassed by, for example, the 0–10m band ($314m^2$) is very much smaller than the area within the 20–30m band ($1,550m^2$). Therefore, the shape of the histogram for the number of birds recorded per distance band (c) is different to that for the 'density' of birds in each band (d). You must combine data from many point counts or line transects to produce these curves, but do not combine those from more than one habitat without careful consideration. Different habitats tend to produce different curves because birds are more or less conspicuous in them – the main strength of distance methods is that they account for these differences.

What the DISTANCE program does, in effect, is to draw out each detection curve and then fit a mathematical model to it. The problem is that there are a number of different models that could be applied and a variety of ways in which the data could be manipulated to ensure better fit of a model estimate. Although you do not actually have to produce a detection curve yourself (the raw data are entered into DISTANCE), we strongly advise that you do this for representative species before using the DISTANCE program. The main reason is that a number of decisions about data manipulation and model fit have to be made, and these depend upon an adequate knowledge of your data set. Producing histograms of detection curves is the best way to do this, and the optimal approach is to produce some initial curves during the pilot study. In this way, you may be able to modify your data collection to avoid some of the problems discussed below.

3.4.2 Shape criteria
Figure 14a shows a good detection curve – it has narrow bands, the number of contacts remains fairly constant over the first few distance bands (the curve has a 'broad shoulder') and there is a smooth and rapidly declining tail. There are a number of reasons why field data may not approach this ideal shape and these are dealt with below.

Figure 14. Bird detection curves – some problems.

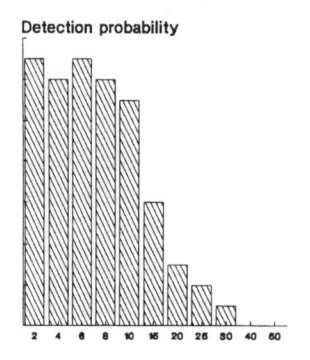

a) Good detection curve with broad shoulder and steep tail

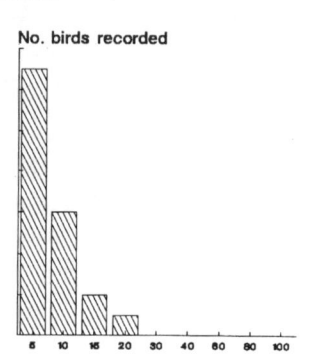

b) Skulking bird often recorded on paths

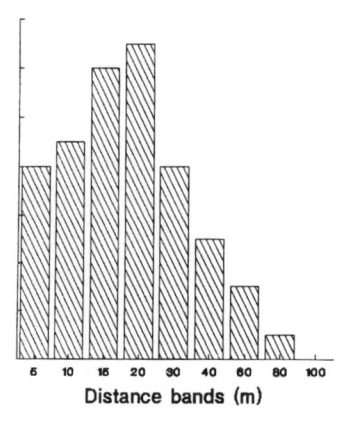

c) Birds move in response to recorder presence

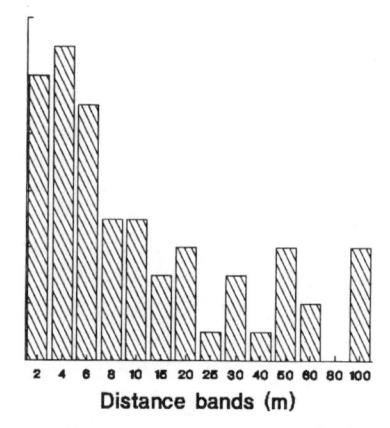

d) Outliers. There may also be a problem with heaping (at 50 and 100m)

Lack of a broad shoulder in the detection curve

This can be caused by missing too many contacts close to the recorder, or by the bird being attracted to the recorder. This is illustrated in Figure 14b and it can have serious consequences for the reliability of the estimates produced. If you have identified the problem during the pilot study you could consider using point counts (if you were using line transects) or use longer point count periods. If you stand still for longer you are likely to detect more of the birds around you and this may produce a broader shoulder for the detection curve for that species. If all the data have already been collected and a lack of a broad shoulder is apparent then you could enter the data in distance bands but manipulate the band distribution and widths to give the best curve.

Higher or lower than expected values at close distances

There can be a number of causes. Birds fleeing from the observer can produce low values for the closer distances and Figure 14c illustrates such a circumstance. Studies using point counts are particularly prone to large fluctuations at these close distances, because the area sampled close to the recorder is very small. This is one reason why you need more data for point counts than for line transects. Altering the widths of the distance bands prior to data entry can produce a better curve and more reliable estimates.

Outliers

Outliers are records of birds detected at large distances from the transect or census station (see Figure 14d). These add little information about bird density and make fitting a model more difficult. As a general rule, outliers should be routinely removed or 'truncated' from the analysis, to enable a better model fit to the data. How much data have to be truncated will depend on the actual detection curve, but 5% of all line transect data is an average figure. A slightly higher percentage (around 10%) of data generated by point counts is usually truncated because there are a higher proportion of detections at larger distances, and these will distort the ideal shape of the detection curve by flattening the tail.

Heaping

If there is a tendency to round off distance estimations to the nearest 10, 20 or even 50 metres, there may be large 'heaps' of records at particular distances surrounded by very few records. This problem may have been recognised and alleviated in the pilot study but if not, grouping the data into different distance bands would help. The first distance band should be narrow and should fall within the 'shoulder' whilst the width of the other bands should increase with distance from the point or transect line.

Cluster bias

This is only a problem if the detection of species is a function of cluster size, e.g. if observations at larger distances tend to be of larger flocks than those close to the observers. Detection distance and cluster size should be independent and drawing some scatter plots and calculation of correlation coefficients will test this. If the observations at larger distances tend to be of relatively large flocks it is a good idea to truncate the data prior to analysis to remove some of the large groups.

Small sample sizes

It can be very difficult to ascertain the shape of a detection curve based on a small sample size (the DISTANCE software will find it equally difficult). One option would be to combine data on the same species from different sampling units, in order to obtain a better detection curve (and eventually a more reliable estimate). This is only valid if there is good reason to suppose that a species will have the same detection curve (or level of conspicuousness) in those different sampling units (for example, the habitats were similar and the surveys were carried out at similar times of year).

3.5 Using the DISTANCE software

An overview of the use of the software and the different options available within the program are given below. Some annotated sample inputs are given in Section 9. For full details, see the program manual (Laake *et al.* 1994).

3.5.1 The basic model

In simple terms, the DISTANCE program 'draws' out the detection curve for each species in each sampling unit and then fits a mathematical model which describes the data. The most important decision to be made is which model to fit to the data. The three main models or key functions are called Uniform, Half-normal and Hazard Rate, and the fit of each can be adjusted by using a 'series expansion'. By default (i.e. unless you tell the program otherwise), the Uniform key function is used because it performs well in a variety of situations.

3.5.2 Alternative models

The basic shapes of the uniform and alternative key functions are shown in Figure 15. The half-normal key function is sometimes used when the level of detection declines quickly over distance. In these circumstances, the data are often not truncated and the half-normal model followed by a series expansion called Hermite polynomial is applied. The Hazard Rate model is more effective for data which show a flat shoulder and long flat tail. The Negative Exponential model with a simple polynomial expansion is occasionally useful for the analysis of poorly collected data. Whatever model is selected, you should ensure that none of the four basic assumptions (given on pages 35–36 have been broken.

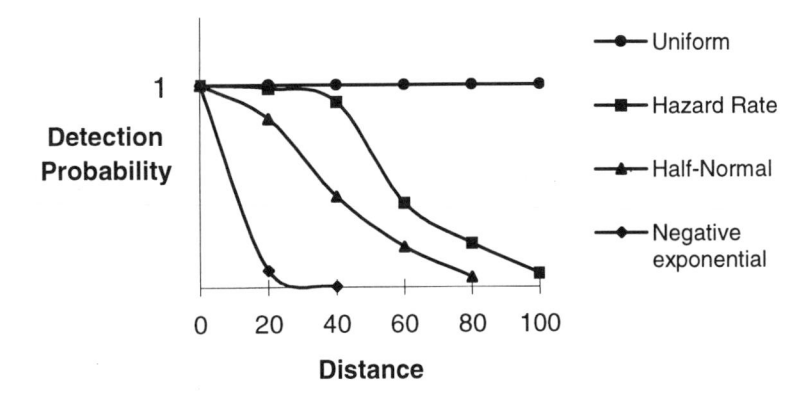

Figure 15. The basic shapes of the uniform and alternative key functions.

3.5.3 Testing model fit

You can fit any of the models to your data but how do you test which is the best fit? Akaike's Information Criterion (AIC) provides a quantitative method for model selection (see Buckland *et al.* 1993). The relative fit of the alternative models may be evaluated using AIC, so that the model with the best fit and least number of parameters will have the lowest AIC value. Therefore, rather than accepting the default Uniform model, you can 'ask' the program to examine the AIC values for each model fit and calculate the density estimates by using the model with the lowest AIC value.

Just because a model is judged to be the best fit of those possible, it does not necessarily mean that it is a close fit or one which will produce a precise estimate. The DISTANCE program uses the χ^2 statistic to assess the 'goodness of fit' of each model. For a number of reasons, it is not a particularly sensitive test, but when you look at the output from the program a significant χ^2 test is a useful warning that the model might be a poor fit and/or one of the four critical assumptions of distance sampling might be seriously violated.

3.5.4 Inputting data

The following is a summary of how to input data into the DISTANCE program. We strongly recommend that you treat this as a very basic introduction and refer to Laake *et al.* (1994) and Buckland *et al.* (1993) before producing your final estimates.

Basically, the data input can be divided into three sections as follows:

Options
Here you describe the parameters of the census method, i.e. whether the estimate is based on line transects or point counts, what the units of area are (e.g. per km^2 or per hectare) and the units of distance estimation (usually metres).

Data
You can actually enter the data here or you can refer the DISTANCE program to another file in which the data are stored. The group sizes and distance to each contact are arranged by sample effort, i.e. per line transect or point count. You also state here how many times each sample was repeated and whether the samples are arranged into different strata – each stratum could be a different habitat. Each sample can also be given a label.

Estimate
This is where you tell the program which model to fit to the data or how it should decide which model to fit. You can also select whether to have a density estimate for each transect, for each stratum or for the whole data set.

Some annotated examples of input files are shown in Section 9. When you try and run the program, it may abort its run for a number of reasons. A common problem is that a group size or a distance estimate is omitted because of an error in data input, so check your data very carefully.

3.5.5 Understanding the output
When the program runs successfully, the results are put into a file which (unless you tell it otherwise) is called 'dist.out'. Much of the output is concerned with fitting the models to the data. The final section contains the population estimates with their standard errors and 95% confidence intervals. Potentially useful statistics in this output are the effective detection distances (effective detection radius (EDR) in point counts). The EDR is the distance from the observer, beyond which as many bird contacts are missed as are actually recorded within the EDR. By comparing the values between species and habitats you can check to see, for instance, if a species is equally conspicuous in different sampling units. If it is and it makes biological sense to do so, you could combine the sampling units and get a larger sample size and (hopefully) a more precise estimate.

Section 4
COUNTING SINGLE-SPECIES

Stuart J. Marsden

4.1 Is a single-species study appropriate?

The impetus to study a single-species or group will usually be a lack of knowledge about, or particular concern for the plight of, that species or group, and/or the funding available from a special interest group. So what is the nature of a single-species study? On the one hand it is a tailor-made and concentrated effort with specific aims and outputs. On the other it may be impractical, unnecessary or a misguided waste of resources. In short, why ignore 99 records of other bird species for the sake of a single record of one species? Consider how the data you collect on the single-species fits into biodiversity thinking – you might find the best area for your species but what about the rest? Single-species studies that can be incorporated into fuller studies may be best in some situations but if you do choose to study just the one species then you must choose a method to suit your bird precisely. The key to doing this effectively is to gather information about your species, its distribution and likely abundance, then fully understand the pros and cons of different census methods.

4.2 Narrowing down the search: information is the key

A complete literature search on the species and its relatives, its habits, habitats and the area to be visited is essential, as is contact with any workers in the field. For globally threatened species, such information has been, or will soon be, documented in Red Data Books covering Africa, the Americas, and Asia. Local scientists may have limited access to such literature but they can be in a better position to contact counterparts in the study area, local community leaders and local hunters. Perhaps a brief visit to the study area to identify possible research sites and methods will be valuable and not too expensive.

It may be important first to look for a species where it was seen last. If the species is present there, useful initial experience of its habits can be gained before searching other areas. If the species is not there any more, then you will still have some important data (without actually recording the bird). Why is it absent (e.g. habitat change, hunting) and which other areas might still support it based on this knowledge? It is important to note that in some species which have suffered declines through over-harvest or through predation by an introduced taxon, the relict distribution may not coincide with its original distribution or reflect the habitat which it most favours.

Figure 16 shows aspects of narrowing down the search for a species. Confirming absence or looking for a species in a likely but unknown area is a sensible starting point. While it would perhaps be a waste of effort to visit an island or region where the species is unknown (Figure 16a), there are exceptions. Some species are known only from a handful of old specimens, some of which can have localities mislabelled. Species can be misidentified in the field, escaped individuals can be recorded in unnatural areas, and every year new species and range extensions come to light.

The next scale down is local presence/absence. In the example (Figure 16b), the species prefers higher altitudes. Just as clear is that the positioning of the sample effort greatly affects the number of birds recorded (A = species extinct! whereas B = common highland species). Sample effort C is perhaps the most successful (C = species common at higher altitude, uncommon at mid-altitude and absent in the lowlands). Local differences in abundance can also be due to rainfall patterns, longitude and latitude, hunting pressure and many other factors, and remember that many tropical birds undertake altitudinal and other local migrations.

The need to look in the right habitat is just as great. In Figure 16c, the species is present within only a small proportion of the area. Again, placement of sample effort is crucial, and in the example there is a need to survey both riverine forest (the species' favoured habitat) and the area of remnant non-riverine forest. Knowledge of whether the species occurs in the remnant forest may be extremely useful in describing its range and habitat needs (is it a riverine forest specialist or is riverine forest just about the only forest left in the area?).

As birds have general habitat requirements, so do they have microhabitat requirements. You need to get right into the microhabitat where your species lives. Again, information is the key: does the species nest in dead trees, does it like the open understorey of primary forest or the closed understorey of disturbed forest? Some information on your bird's microhabitat may be gleaned from field guides or experienced birders. If specific information is not available for your species, then details on related taxa may help. Remember though, that available details of microhabitat may not represent the species' true requirements, rather the habitats in which it is most easily detected or, again, the only habitat left.

Figure 16. Narrowing down the search for a species.

a) Regional presence

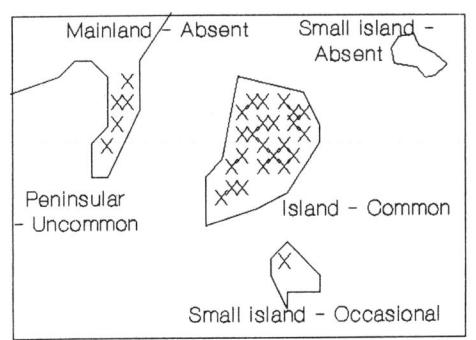

```
         Mainland - Absent    Small island -
                              Absent

         Peninsular
         - Uncommon           Island - Common

                    Small island - Occasional
```

b) Local presence

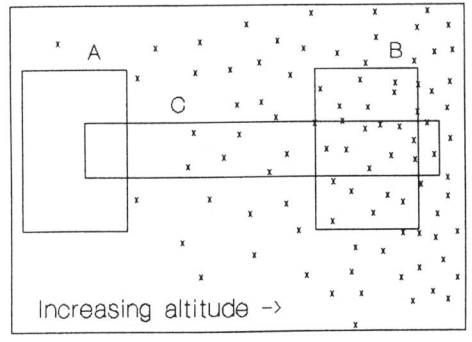

```
         A              B

              C

         Increasing altitude ->
```

c) Restricted habitat

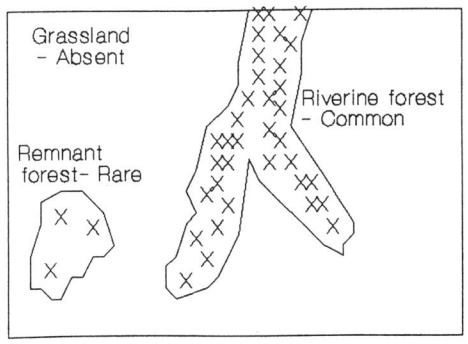

```
         Grassland
         - Absent
                              Riverine forest
                              - Common
         Remnant
         forest - Rare
```

(x = *species recorded*)

Time of day is obviously an important consideration in survey design (Figure 17). Some 'windows of study' are fairly obvious, e.g. nocturnal owls are best looked for during the night! Others are not so obvious. Nocturnal birds may be sought (in their active phase) at night but searches for roost or nest sites during the day may be just as important (and these require different methods). When we take account of seasonal time factors, windows of study can become quite complex (Figure 18). There may be a specific time of year during which birds sing (some tropical birds may sing extremely infrequently), or a certain time of day. The timing of your fieldwork and the methods you choose should reflect these considerations.

Figure 17. Time windows for the study of a resident diurnal forest bird using a distance method (the species sings between July and September). ✓= good time, ✓✓= very good time.

Time/ month	J	F	M	A	M	J	J	A	S	O	N	D
Dawn	✓	✓	✓	✓	✓	✓	✓✓	✓✓	✓✓	✓	✓	✓
Morning	✓✓	✓✓	✓✓	✓✓	✓✓	✓✓	✓✓	✓✓	✓✓	✓✓	✓✓	✓✓
Midday												
After-noon	✓	✓	✓	✓	✓	✓	✓	✓	✓	✓	✓	✓
Dusk	✓	✓	✓	✓	✓	✓	✓	✓	✓	✓	✓	✓
Night												

Time/month	J	F	M	A	M	J	J	A	S	O	N	D
Dawn	4	4	4	4	4	4	4	4	4	4	4	4
Morning	3	3	3	3	3	4	4	3	3	3	3	3
Midday												
Afternoon	3	3	3	3	3	4	4	3	3	3	3	3
Dusk	4	4	4	4	4	4	4	4	4	4	4	4
Night	1	1	2	2	2	1	1	1	1	1	1	1

Figure 18. Appropriate survey methods, time of day and month for a (mostly) nocturnal owl which calls between March and May, and which has fledglings in June/July.
1 = transect searches, 2 = playback of call, 3 = roost-site searches, 4 = search for active adults/juveniles.

4.3. The nature of the beast: rarity

There are many forms of rarity. Absolute rarity means that numbers of a particular bird are known to be low. With a wild population of only one, Spix's Macaw is obviously one of the world's rarest birds. The absence of actual population figures for most bird species means that relative rarity is often used. Thus, Species A is rarer than Species B, or common in one region or habitat but rare in another, or rarer than it was 20 years ago. The above terms for rarity are valid but some instances of perceived rarity are not. Some species may be difficult to find for several reasons but this may be very different from actual rarity. Some little-known species are seen as rare because previous expeditions have looked for them in the wrong place or used the wrong methods. There may be a tendency to look upon a species that has not been seen in the wild for many years as rare: has anyone actually been to look for it?

Rabinowitz (1981) described three components of actual rarity: small global range, restricted habitat and low population density. Some species can 'suffer' from more than one of these, the worst case being of a species with a highly localised range, within which it occurs in a very specific habitat and, even in this habitat, it occurs at low density. These are natural ecological patterns of rarity but they have serious implications for the bird surveyor.

One of the forms, small global range, has been introduced in Figure 16 so let's presume that you are 1) in the right place, and 2) in the right habitat (the habitat of a rare species may be poorly known, so you are looking in a range of possible habitats in which it may occur).

Figure 19 shows some different patterns of distribution/rarity. For a certain 'survey effort' (which could be one person looking for one month), the rarer the species, the fewer records the expedition will amass. Sample efforts A and B are superimposed on the figures. In the case of a territorial bird (19b), individuals tend to be fairly evenly spread out. Note that the position of the sample effort does not make such a difference to the number of birds actually recorded. Increasing the sample effort (say from one to two months) may roughly double the number of contacts with individuals. In the colonial or clumped system (possibly the result of restricted habitat), the situation is very different: (B) records many birds while (A) records none. In general, the more clumped a species' distribution is, then the larger the area that the sample effort must cover to get a true idea of the average abundance of the bird.

One type of 'rarity' which is important in bird surveys is that some species, for various reasons, can be difficult to detect (they can be nocturnal, cryptic, or they can be disturbed easily). For these species (19d shows a cryptic, uniformly distributed one) the problem is being able to record the individuals which are actually present. While it may be best to cover a lot of ground to census colonial birds, for cryptic species it may be better to concentrate on a smaller area and make sure that you search well enough to find most of the birds present. It is very important to understand how your species fits into these patterns of rarity.

In a multi-species survey, it will usually be unwise to jump from distance sampling to another technique for the benefit of just one or two rare species. Alternatives could be to stay longer, or to devise specific/focal studies for the species as an aside to the distance sampling regime. In single-species studies, the jump can be just as drastic, so the important thing is to decide fairly early (during the pilot study) whether the standard method is going to be appropriate. You should try a distance method and extrapolate the number of records accumulated over the first days to the number you can expect during the whole fieldwork period (see Section 2). Can you restrict the survey to the types of habitats where the bird occurs?

For various reasons, you may not be recording the bird sufficiently often. If it is a cryptic species, expanding the count period may work. Alternatively, the bird's distribution may be so clumped that you are missing the aggregations or colonies. In this case you need to locate the aggregation and make a total count or estimate of each aggregation. You may be able to locate the species more easily when it flies by looking over large areas of forest from a vantage point. In cases of extreme rarity you may have to use all your birding skills just to find it, or ask local people. Contact with local people concerning your species and the project you are doing can be extremely valuable (see Section 4.6.5). In other cases, you may have to search specifically for a species and concentrate on its habitat associations (see Section 6).

Figure 19. Effect of distribution and rarity on survey results.

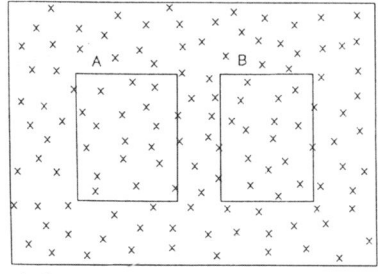

a) Common species

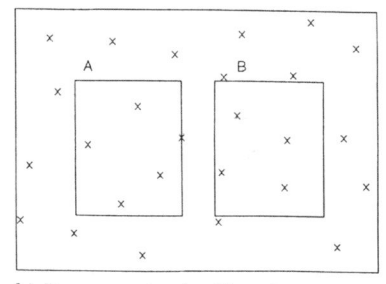

b) Rare species (uniform)

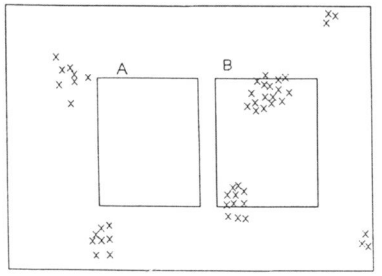

c) Clumped

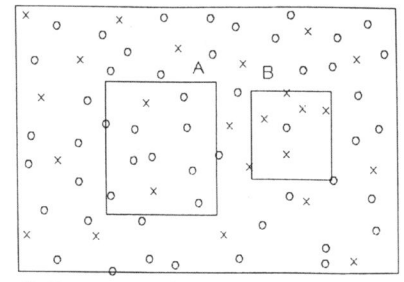

d) Cryptic (and uniform)

x – bird recorded; o – bird missed

4.4 Problem species and difficult habitats

Birds come in all shapes and sizes and their habits and habitats do not always make counting them straightforward. This is not to say that unusual birds are always more difficult to work with than 'ordinary' ones. In fact, some counting methods actually use the unusual characteristics shown by their subjects to their advantage. Again, information about the natural history of your subject species is crucial in tailoring methods to suit the situation. Below are listed some special characteristics shown by birds, some examples, and the implications of these for censusing. This is followed by more specific methods to deal with problems that may arise.

4.4.1 Bird colonies/aggregations

The distribution of many bird species will be clumped to a certain extent. The more clumped, the bigger the gaps will be between bird occurrences (Figure 19c). Clumping can be very pronounced in a number of situations:

Many birds breed in colonies although they may disperse in the non-breeding season. Colonies may range from just a few pairs to many thousands of birds. In the former case (e.g. a small colony of starlings or mynas in a large dead tree), normal census methods may be suitable, but in the case of single large colonies, specific searches followed by total counts would be better. In these cases, information from local people can be crucial and not too difficult to obtain if colonies are spectacular, or if birds, eggs or nests are harvested. The focus for colonies may be obvious (e.g. caves for swiftlets or oilbirds; cliffs for parrots and hirundines; hot springs or beaches for megapodes; riverside trees for waterbirds) or less easy to predict ('traditional' nesting grounds for megapodes, or individual trees for weavers).

Many birds will roost in groups for at least part of the year. As with breeding colonies, both the size of roosts and the focus for the roost (large trees, caves, cliffs, mangroves, or mudflats) varies considerably. Again, prior knowledge of where the species may roost and specific information from local people will be invaluable. In this case the fieldworker has some useful options; birds can be studied away from the roost (using normal census methods), as they fly to/from the roost, or actually at the roost. Birds will often roost by night, but there are other cues, including tidally-driven roost patterns and periods of post-feeding 'roosting' in frugivorous birds.

Some species, such as manakins and birds of paradise, aggregate around lekking grounds, to display to potential mates before breeding. Such aggregations are usually quite small (fewer than 50 birds) and subject birds may be counted as clusters with distance methods. Leks can vary in their conspicuousness to the recorder and in their timing (e.g. early mornings at certain months of the year). Local people may know of lekking sites (they may collect birds or colourful feathers). Some lek sites can be confirmed 'out of season' because birds alter the local vegetation (clearing the ground) or leave signs (feathers or faeces).

The above cases of aggregation can help the fieldworker to count them. However, many birds (such as pigeons, finches, tanagers and jays) simply travel and feed in large groups. Consequently, you may go a long time without recording any and then stumble across too many to count (i.e. encounter rate low and group size unknown). In all but the most extreme cases, this problem is best tackled using standard methods but the following must be considered. First, you should make sure that you have enough bird records – not the total number of birds seen but the number of groups encountered. Second, you must be able to predict the number of birds in each group – every time you see a group (on census or not) try to estimate the group size. You can then substitute the mean group size of these encounters for those during the census where the group size could not be determined.

4.4.2 Cryptic and understorey birds

For a given population density and search effort, the detection rate of a species will depend on its conspicuousness. Many of the world's most beautiful and elusive birds are cryptic understorey or ground dwellers. Remember, a birdwatcher usually aims to *see* a bird, whereas the bird surveyor can record it by sight or sound. Knowing a bird's call with confidence can increase its encounter rate tenfold or more. Get this information from literature, tapes, previous visitors, local people, or first-hand during the pilot survey. Once this is known, as long as it is not too rare, then the species can be censused using a variation of VCP method (section 3.3.6). Alternatively, if the species flushes easily (e.g. gamebirds and many other ground-dwelling birds), then walking transects may be better. Your aim could be to count the birds as you flush them (while trying to ensure that you don't count them twice).

4.4.3 Canopy species

Many rainforest birds use the upper storeys (30–70m), making their detection difficult (especially as some move around quickly in mixed species flocks). There is no easy way round this problem of detection, although a variant of distance sampling may again be the best compromise (section 4.5). In a nutshell, the answer is to find out the proportion of birds actually in the canopy that you can record from the ground (in distance sampling, the probability of detecting birds at zero metres <1). Perhaps the most feasible way of doing this is for one team member to do a normal point count, while several other recorders stand around him/her looking into the canopy through gaps and trying to record every bird that is there. If this is done many times then density estimates from the single observer can be corrected, by relating them to numbers of birds seen by the other observers. For example, if the single observer records five birds per point count and the other observers record ten between them, then estimates should be doubled. Of course, this is a very approximate method and in some cases it may be near impossible to know the proportion of birds you are missing. Perhaps the most important point to appreciate is that the proportion of canopy birds that you detect in a forest with a low canpoy may be much higher than in tall forest.

4.4.4 Mixed-species flocks

Many species join mixed-species flocks for at least some part of the year. It is important to distinguish between those, mostly insectivores, which move around, from feeding aggregations such as several frugivorous species at a fruiting tree. In some respects mixed-species flocks can be seen as a variant of single-species flocks (section 4.4.1) but with the added problem that both species-composition and group size are unknown. Flocks may be rarely encountered, move very fast through the forest, and contain variable numbers of individuals and species. For an estimate of abundance for one of its constituent species, we need to know (1) the number of flocks in a given area, and (2) the presence and number of target species within each flock.

4.4.5 'Aerial' birds

Although birds can spend between 0% (flightless) and almost 100% (swifts) of time in flight, most spend less than 50% of their time flying. Also, in forests, most birds do their important 'business' (feeding, breeding, roosting, etc.) when they are not flying large distances. For the great majority of species, a record of a perched bird is much more important than one in flight. In most species, flying birds can simply be omitted from distance sampling estimates. In a minority of very mobile birds (or ones which are cryptic at rest and conspicuous in flight), however, specific techniques may be the only/best way to count them. These include swifts, swallows, some raptors, etc.

4.4.6 Nocturnal and crepuscular birds

Birds which are difficult to count during the day are obviously a special case. Distance methods will probably be unsuitable for fieldwork at night (not least for reasons of safety). Spot or territory mapping of calls, encounter rates along transects, or simple presence and absence in different habitats may be the only way of collecting meaningful data. Marking trees with brightly-coloured tape may allow relocation of sites during daylight. Remember that many nocturnal birds prefer edges or clearings so transects which proceed along paths or roads (rather than randomly through forest) may overestimate bird abundance. Transects could work if roosting birds or nests are looked for during the day (several people could walk side by side, blanket searching for likely cavities or roost sites).

4.4.7 Other 'problem' birds (waterbirds, birds of prey, migrants etc.)

Some tropical waterbirds can be counted at aggregations but others occur along watercourses within forest (e.g. forktails, kingfishers, ibises, herons). In some respects this makes them easier to count (their habitat is almost two-dimensional). Line transects along streams may yield density figures of bird pairs per kilometre of stream. In some cases (e.g. forktails) spot-mapping of territories works as birds may flush to the edge of their territory and fly behind the observer. Nests of stream birds may also be relatively easy to find.

Some birds of prey are very difficult to census using distance methods. Here the answer may be to count the birds as they fly above the forest or to spot map their nest-sites. Finally, some birds are either only known from migration, or can only/best be counted as they migrate. Counting large birds such as cranes or raptors at migration bottlenecks is similar in many ways to counting birds flying to/from roosts.

4.4.8 Special habitats/niches

Any bird's niche or lifestyle will seem special if you look hard enough at it, but in conservation studies, some will appear more specialised than others. A bird which occurs mostly in, say, mangrove, riverine forest or bamboo, must be looked for mainly (but not exclusively) in that habitat. In many cases, the census method used can be exactly the same as for general habitats. In others, such as mangrove, variations such as the use of a boat require only common sense, as long as the effect of the boat on birds is considered. Other habitats such as particularly steep terrain or montane areas (where bird density may be very low) may need specific consideration.

Specialisation may be behavioural or resource-based. In the neotropics a number of species are closely associated with swarming ants. A few African/Asian species are associated with bee nests (honeyguides) or with particular events (e.g. bee-eaters and raptors with bushfires). Some of the above species can be counted with standard methods, but others are rare and so specialised that studies must be focused specifically on the resources that limit their distribution.

4.5 Tailoring distance sampling methods for individual situations

This section describes how distance methods can be tailored to suit individual species types. Of course they are only broad recommendations. It cannot be stressed too much that for your census to be successful, you should take time to understand the principles of distance sampling, and how the behaviour of your species fits in with its assumptions (see Section 3). To recap:

- your plots or transects should be positioned so as to form a representative sample of the area under study;
- the probability of detecting birds on the transect line or point must be certain;
- birds must be counted at their initial location, prior to any natural movement or movement in response to the recorder's presence;
- distances to bird contacts should be known accurately.

'Ordinary' birds

Count period: Usually 5–10 minutes in multi-species surveys. For a single-species survey, 5 minutes may be long enough.

Search effort: Normal (see Section 3).

Flying birds: Ignored/omitted.

Spacing of points: For VCP method distances of 200–300m between census points is usual. You must find the best compromise between unnecessary walking and the risk of a bird being counted very close to one station and then of it moving to a position very close to the next station.

Cryptic birds and cue counting

Count period: Should be longer than for the 'ordinary' species. Perhaps 10–12 minutes. This is to ensure that high proportions of the birds close to the recorder are actually detected. A very different approach, which may be useful when surveying gamebirds and other ground-dwelling species (especially in grasslands), is to walk transects and to count birds as you flush them. This is a variant of the standard transect method, where the aim is to flush as many birds as possible from on, or near the centreline (without double-counting them).

Search effort: Concentrate on visual and aural cues within 20–30m of the central point. Because the count period is long, be careful not to record birds which you think may have entered the plot after the count period began. Some cryptic birds can be disturbed easily so approach the station very carefully and record any birds that flush due to your arrival as being present at the station. At the end of the count period, it may be worth walking around the plot or 'pishing' to make sure you record stubborn birds.

Other: Stations may be positioned fairly close together (perhaps 100m minimum). Dawn and dusk may be the best time to census many cryptic birds.

An alternative approach is to use a variant of distance sampling called **cue counting** (see also Section 4.6.3). Perhaps the most important 'cue' in the census of a cryptic bird species is its call (the species could be a partridge, pheasant or pitta, for example). The cue counting technique has two components:

1. **Estimating cue density**: Sampling the 'cues' using a standard VCP method or a variant of a distance transect method. While in normal distance sampling, the fieldworker records an individual bird only once per census plot; in cue counting, he/she records every time the cue is given (i.e. every time the bird(s) call(s)). For example, during a ten minute point count, there might be three calls (cues) given by an unknown number of birds within 30m of the recorder.

2. **Estimating cue rate**: This involves estimating the number of times that an individual of the target species gives the cue during a given time period. To do this, you will have to perform focal studies on several individual birds for several hours to find out an 'average' call rate for birds (see also Section 4.6.3). Take care to ensure that minimal bias creeps into your estimate of cue rate. For example, you should estimate a species' calling rate during the same time of day as your census will take place (as birds will call more at some times of day than others). Also, call rate may depend on the density at which birds occur, so undertake focal studies of calling birds in several different areas.

An option is available in DISTANCE to deal with cue count data. First, TYPE=CUE must be chosen. Then, the cue rate must be entered using the CUERATE option. If points are used, then you should express the cue rate as the number of times that the cue is given per count period. Thus, if your count period is ten minutes per census plot, then the cue rate will be the number of times the call is given per ten minutes (e.g. if the bird calls six times per hour, then cue rate equals 1, for count periods lasting ten minutes). Further information is given in Buckland *et al.* (1993) and on pages 24 and 34 of Laake *et al.* (1994). One complication is that since more than one cue counted at a census plot may come from the same individual, then the distances entered are not independent events. Because of this, estimate variances should be calculated using the BOOTSTRAP command.

Aerial Species

Count period: Must be near-instantaneous as birds are extremely mobile. Count periods lasting any more than a second or so will seriously overestimate bird density.

Search effort: The recorder looks directly upwards and estimates the horizontal distance to bird contacts at that moment. Remember that birds will be much easier to detect in open areas than in closed canopy forest. An alternative and possibly useful method is to look up through binoculars or even a standard sized tube. This will restrict your view of birds to a standard air volume. This method will not produce actual density estimates but may still allow comparison of encounter rates between habitats or areas.

Other: Census points can be positioned very close together. Each instantaneous count is then entered into DISTANCE as a replicate of the one plot.

Parrots, hornbills, toucans, etc.

Count period: Ten minute counts may be necessary to increase likelihood of recording birds which can be extremely cryptic at rest and obvious when flying. Be careful not to record birds which fly into the plot during the count.

Search effort: Concentrate on perched birds within 50m of the recorder. At the end of the count period check for stubborn birds (see 'Cryptic birds'), particularly ones which may be sitting silently in the canopy or in nest holes. If this is to be done, then it is important to standardise this period of flushing as much as possible, both between census stations and between different recorders. Parrots may respond to your presence: they may give alarm calls, stay silent, fly away, or even fly towards you. Remember that you must record birds in their initial positions.

Other: Many such species are rare, so many stations are needed to produce good population estimates. I prefer the VCP method to transect methods for parrots. The main reason is that with point counts, there is a better chance of recording all birds close to the observer. If transects are chosen, then they must be walked very slowly and carefully (bird detection at distance zero metres is paramount). Stations can be placed 200–300m apart in forest, but maybe 500–600m in very open habitats. Flying birds must be ignored in density calculations (except those which fly out of the census plot). The best time of day for parrot and hornbill census (using VCP method) is often between one hour after dawn and 1030h (and perhaps between 1500h and an hour before dusk).

Large groups and mixed species flocks

Count period: Flocks of birds may be fairly conspicuous and move around quickly. Short periods (2–5 minutes) are appropriate. This does not mean that you cannot spend time after the count period has finished to identify species and estimate group size (although birds must obviously be recorded in their initial positions, not where they were when you identified them).

Search effort: For bird groups, you can either estimate the distance to the centre of the flock (and then enter the data into DISTANCE as 'clusters') or estimate distances to individual birds (Section 3).

Difficult habitats

Count period: In areas of low bird density, the answer is *not* to increase the count period but to increase the number of stations or kilometres of transect walked.

Other: In some situations, transects may be the only way of gathering enough bird records. They may be appropriate in some montane areas, where trees are not as tall as in lowland forest (birds may be easier to detect). However, surveying while walking may be dangerous, and you must make sure that the habitat through which your walk route passes is fairly representative of the forest as a whole. In some situations, your view at a station may be obscured by cliffs. There is an option in DISTANCE to account for this. For example if at one station you can only see half the plot (i.e. 180°), then you can enter this point as a 'half plot'.

4.6 Specific techniques for special cases

4.6.1 Spot mapping

Spot mapping involves plotting the position of birds on a map. The technique has been used primarily in temperate regions to count the territories of singing males. The restraint is normally that it is very time-consuming. You need 6–10 visits to an area and these visits must be sufficiently spread in time as to be independent. Although little-used in the tropics, the advent of GPS technology may make spot mapping fairly useful in some situations (although it will not usually be accurate enough for use in closed-canopy forest). Uses of spot mapping include territory mapping of singing males and mapping nests, colonies, roosts etc. They may be suitable when a species is too rare for distance sampling, or when forest patches are so small that all the birds can be counted and may work well for species such as thrushes and territorial flycatchers. They may not be appropriate outside the breeding season, for birds with asynchronous breeding seasons, for birds which do not hold strict non-overlapping territories, or for birds which sing in more than one area in a single season. Little or no data are available on the mating systems of many tropical birds, so you may be left with 'unknowns' which could jeopardise the value of your population figures (e.g. there may be many non-breeders, especially in populations of long-lived species). Likewise, counting singing males alone may not provide the same index for all species in all circumstances, when sex ratios are biased.

The method is time consuming. You will need to locate every individual in the area, or use a calibration technique to predict the likely numbers that you have missed. How much time or how many repeat visits it takes to find the birds will depend on its ease of detection (e.g. strength and frequency of song, conspicuousness of nest). There are several variations on the basic method which may be appropriate in some cases. Playback of tape recorded calls could be very useful, particularly in surveys of nocturnal owls. A useful technique for some species is to flush birds to the edge of their territories (making territory definition much easier). With playback and flushing techniques, it is very important to minimise disruption to birds.

4.6.2 Counting birds at or near aggregations

Counting birds at roosts, colonies, leks etc. will usually (but not always) involve direct counts of all the birds present. You may not be able to count all birds, either because there are too many, or because you cannot see them all. In cases of very large numbers, you may have to sample the population. In a cave holding many birds, could you derive a measure of birds per m^2 by some cunning means? (e.g. counting the birds within the beam of a torch held a known distance from the cave's wall). Another possibility might be to mark out known areas on the cave or cliff side and then count birds within these areas (marks could be used for several years). A different approach is to select (many) birds at random and estimate the distance from each to their nearest neighbour. Counting large bird roosts in trees is difficult. Maybe select a few trees, count the birds and multiply up. Can you photograph all the birds for counting later? A useful method may be to set up a telescope, binoculars or even a tube, and count the birds visible in the field of view. Then calculate the area of your sample field of view and relate to the total area. In some cases, it may be better to count bird signs than the birds themselves. In busy colonies, active nests or nest holes may be an appropriate sampling unit.

Bird aggregations take many forms, as will the most appropriate methods to count them. There is great scope for devising new and good methods for individual cases, so think about it hard! Apart from some considerations in presenting your results, there are a few specific considerations. First, can you count the species in its aggregating and non-aggregating phases? Ideally, you could count a species (e.g. a parrot) during the day using a standard method, and then in the evening, at or near its roost. Second, are you sure that you have located all the aggregations in the area, or the proportion of individuals which join these groups? Third, have you adequately accounted for variability in counts? You may need to take the average of several fieldworkers' counts, or count the aggregation on several different occasions (times and dates).

The other way to count roosts is to count birds as they move to or from the roost. In many respects this is similar to counts of migrating birds such as birds of prey or cranes at 'bottlenecks'. Figure 20 shows a possible method. It is helpful, but not absolutely necessary to have more than one fieldworker counting at one time. The distance from the observers to the roost will also vary: in general the bigger the roost, the further away the counters may have to be, so as not to be overwhelmed by bird records. You can attempt to cover the whole circumference of the circle (B), or just a sample of it (A). The latter may work better as double-counting birds can be a problem with the former regime, and will seriously overestimate roost size. Using (A), the proportion of the whole circumference which is covered needs to be known. Remember that some birds may enter/leave the roost in the dark. It is also unlikely that departure or arrival directions will be uniformly spread.

Figure 20. Counting birds flying to/from roosts.

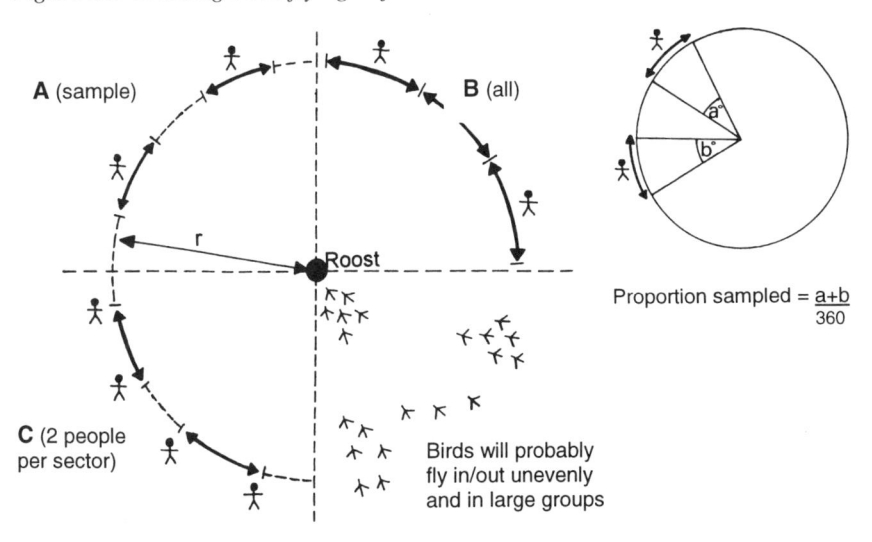

A (sample)

B (all)

Proportion sampled = $\frac{a+b}{360}$

C (2 people per sector)

Birds will probably fly in/out unevenly and in large groups

4.6.3 'Look down' methods from vantage points

Whereas most census techniques are 'look up' methods, there are useful methods which involve looking down on forest from vantage points or aircraft. Aerial surveys are used for waterfowl and many other birds (although not often on expeditions in the tropics). Long watches from hillsides, cliffs and large trees have been used in many studies (particularly by The Peregrine Fund, USA), both to enumerate populations and to investigate raptor behaviour and habitat associations. With look down methods, you usually need to know the area of your survey. In discrete forest blocks or valleys this can be gleaned from maps, but in other areas you may have to work out the area using trigonometry (take compass bearings to points on the edges of your site and draw your own map). Below are two examples of their use and suitability:

Surveys of parrots in the Caribbean

Parrot populations on Caribbean islands have been monitored for many years using long watches. The method could also work for some birds of prey, pigeons, toucans and hornbills. Observers overlook forested valleys and plot each birds' flight path. In some cases, two or more observers are positioned in different places around the valley/area, and these note the exact timing of each flight: a picture of movement can then be developed for bigger areas. This method has produced some very precise results, but may only work well with conspicuous birds which fly above the canopy, in areas with suitable vantage points and where forest occurs as fairly discrete blocks. Most important, it may only work with rare species – where there are so few birds flying around that you can be sure of following individual birds. The method also allows area usage and habitat association studies. For example, what proportion of all birds flew to, or from, primary forest (rather than areas with scattered trees)?

Cue counting from vantage points

In cases where birds are not extremely rare, it is often difficult to identify individual birds with confidence. However, watches can still yield population measures and these may be useful in very rugged terrain such as steep valleys (where look up techniques are difficult). The method can have one or two components:

a) Within a known area, the number of flights made is recorded and expressed as mean number of flights per hour per unit area (a relative abundance index). The assumption is that the more birds there are, the more flights you will record. Remember, most birds tend to fly more early in the morning and in the evening than in the middle of the day. A serious limitation is that birds in one area may fly more than those in another. This could easily be habitat-dependent, with birds in poor habitat perhaps having to fly further.

b) An extension of this method and a possible way round the above problem, is to find out how much time an average bird spends in flight per hour. From your vantage point, locate a bird in flight and follow it until it perches. Then record the amount of time it spends sitting/feeding. When it takes off again, time its flight. Obviously, you will need to do this many times.

So now you know (1) what fraction of its time the average bird spends in flight, and (2), how many flights are made (by an unknown number of birds) per hour within a known area. To work out your population estimate, divide (2) by (1). For example, an 'average' bird spends 1 minute out of each hour flying. You recorded ten minutes of flight by all birds in one hour. Your population estimate in the area is ten birds. The essential parts of this method are that you, (a) detect every flight made by all birds in the area (so several observers may be necessary), and (b) your data on how much a single bird flies is good enough. This form of censusing is cue counting and in this case, the cue is bird flight. The cue can also be bird calls (see Section 4.5 on cryptic birds for an example of cue count distance sampling).

4.6.4 Nest searching/counting limited resources

Many of the world's threatened bird species have specific habitat requirements which can make them sensitive to habitat change and make their distributions very patchy. Some may be counted using standard techniques but in others the census must focus on their most important or most characteristic habitat association. Parrots, along with many other birds (such as large woodpeckers) nest in cavities in very large trees. They can also be very rare, and one reason for this may be that very few large trees remain in certain habitats. Therefore, a focus for bird census could be nests, firstly because nest availability may limit population size and secondly because nests are characteristic and so are relatively easy to find. Similarly, many birds nest in dead trees and again nests, rather than birds, can form the sampling unit (see Figure 21).

Nests themselves can be characteristic, but so can bird behaviour around that nest. Forest birds of prey can be rare, and they can range widely, making census (from vantage points) difficult. If, however, the sampling unit is the nest, rather than the birds themselves, then characteristic behaviours (returning to nest, displaying and calling) can be used to calculate the number of pairs in an area.

The limiting resource, or at least the focus, for birds associated with ant swarms may be swarms themselves. It may make sense to count these resources: for example spot mapping or counting swarms along transects. A consideration here is that swarms may be much commoner in certain habitats (e.g. primary forest) so stratify your sample. Another consideration is time of day, with swarms being active in the middle of the day, rather than morning when the majority of birds are most active. Once you know the frequency of swarms, then you need to focus on the presence/abundance of your bird species at those swarms, using direct observation. Perhaps the best method for censusing antbirds is to use standard methods in the morning and then focus on swarms in the middle of the day. Variants of the above method may be appropriate in other cases; parrots at clay-licks, honeyguides at bee nests, or birds which nest in holes in banks.

Figure 21. Counting nest trees (or other limited resources).

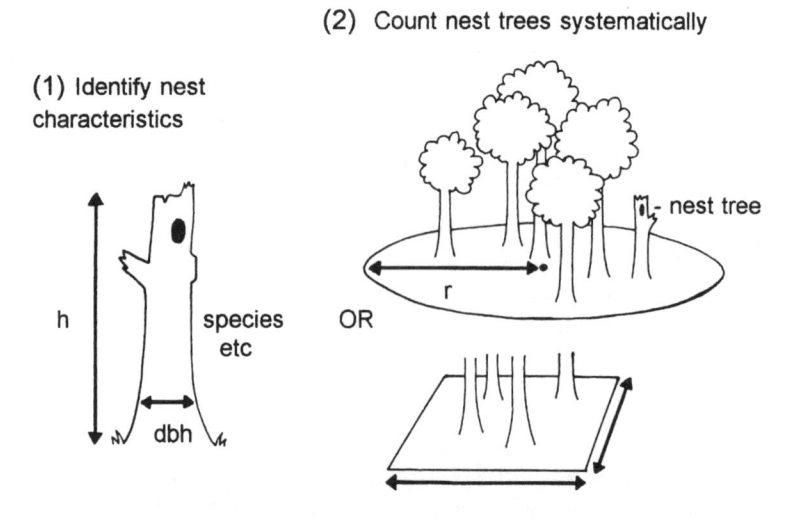

(1) Identify nest characteristics

h, species etc, dbh

(2) Count nest trees systematically

nest tree, r

OR

(3) Calculate number of nest trees per unit area

e.g. 10 nests per km^2 in primary forest
2 nests per km^2 in agriculture

4.6.5 Interviews with local people

It cannot be stressed enough that information from local people can be an essential part of bird census. In the initial stages, local people (who know the forest) can lead you to likely study sites and extremely rare birds. Illustrations of target species will help with communication but remember that local people will recognise many species by their voice or habits, rather than specifics of their plumage.

As the survey gets under way, information about relative abundances can be gained from careful but informal interview. Is the species commoner in area A than B, in one month more than another, or was it commoner ten years ago? Local hunters can give useful information about a bird's ecology, particularly its nest sites, breeding season and favourite foods. Interviews are a particularly good source of information on birds which are economically important to local communities. Some birds will be pests, while others may be traded, kept as pets or eaten. In the latter cases it is crucial that you keep the discussions both amicable and neutral. Don't talk about global declines and cruelty, or you may get either no information or, worse still, information that is aimed at keeping you happy!

If interviews are well thought out, then you may be surprised by how much accurate information on bird capture/trade can be gleaned, even if that trade is illegal. It might help not to aim your questions at the interviewee (are birds still caught in other areas?). Bird prices are easy to find out, and how they change over the years might give useful clues as to population trends. Don't assume that if you are told that many birds have been captured from a certain area that there will be few left there: the opposite can be more likely. Bird markets are a reasonable source of information. Again, keep it amicable and remember that the further you are from the forest the less reliable your source of information may be (particularly as to where birds have come from).

Finally, information given by several different sources is much more reliable than that given once. The most successful interviews are those which ask the same simple and unambiguous questions to many different people. Many more details about planning, implementing and interpreting the results of interviews is given in *Expedition Field Techniques: People Oriented Research*, published by the Expedition Advisory Centre (see References, Section 8).

4.6.6 Mark-recapture/banding/home ranges

Catching birds (usually with mist-nets) has many uses in long-term ecological studies but has limited use in short census studies. There may be some advantages in being able to handle some birds (e.g. nightbirds, very cryptic or difficult to identify species), to find out if birds are breeding, or recognise individual birds (banding/marking) but bird-catching is usually too time-consuming to be considered in the majority of surveys. It also requires intensive training. In some cases it can be worthwhile, for example, capture rates of birds per metre or m^2 of net can be a useful technique for deriving indices of abundance for understorey birds and can also be used for diversity studies. If bird capture methods are used, a scouring of the vast amount of available literature is an absolute necessity, both to get the method right and to avoid harming the birds.

4.6.7 Scientific birding

I had quite a heated argument with someone once over the status of a little-known bird: I was convinced that it was "fairly common in places" whereas he was adamant that it was "rare to locally absent". Without quantification, such assessments cannot be compared, and many studies for which the end result is such a description deserve criticism. But some situations are incredibly difficult: your species might be nocturnal, cryptic, have no characteristic habits and be extremely rare (even local people may not be familiar with it).

It should be apparent fairly early on in the single-species study whether distance sampling or other techniques will not work (this may not be so obvious in a multi-species survey). The species may be so rare that the fieldwork turns from census to search, trying to find at least one record of the bird (i.e. is the species extant in the area?). To get that record, the random sampling approach so necessary in census should be replaced by a specific search in the most likely areas/habitats. This change of approach (i.e. to birding) should not mean the loss of all systematic recording (Section 5).

The calculation of encounter rates is better than just saying we recorded the bird four times. Your data becomes an encounter rate (ER) of 0.04 per hour of search (if you looked for 100 hours). Better still is stratifying the sample by habitat: ER in primary forest = 0.06; ER in logged forest = 0.02. This is the start of a repeatable method: describe your method even if it doesn't seem much of a method! For example, did you start looking just after dawn, or were you restricted to later in the day? Did you have to stick to paths? Did you know its call or were you relying on actually seeing one? Which habitats did you sample? Remember that the habitat in which you didn't record the bird can be as important as the one in which you did record it. Compare the above ERs with ER in primary forest = 0.08, ER in logged forest = 0.00. Your lack of records in logged forest may have serious implications for assessing the status of your species – but only if you describe your data properly.

4.7 Interpreting and presenting results of specific studies

Whether you tailor existing methods to your situation, or you devise your own method, you must describe your technique completely. Things that may seem obvious to you (because you did them) may not be at all clear when someone repeats your work in 50 years time! It may be important to show the locations of vantage points, roosts, colonies or other features on maps (with coordinates). Report how many fieldworkers collected the data, what time of day it was collected, and anything else which will make the repeat census more comparable with your own.

Likewise, your study may itself be a repeat survey. In this case, the method used in the past may be the best one to use. If so, then there is no problem and you must follow their methods as precisely as possible. In other cases, however, you will be able to make improvements or devise totally new methods. Do this, but remember that bird population monitoring depends on comparability. Is there scope in your study to use both the previously used methods and your new ones? This has an obvious advantage: you get a population figure that, although possibly wrong, is comparable with the old one, along with a new (and hopefully more accurate) one which can form the baseline for the future.

As introduced in Section 2, there are two components to a population estimate; its accuracy and its precision. In distance methods, the reliability of results are, to a certain extent, computed statistically. If you devise or tailor your own methods, then errors can be harder to quantify and yet are even more important. For example, in a roost count, did different observers' estimates match? One person counted 100, while others reckoned 90, 80, 90, 110 and 130. So which estimate do you use? In this case, the mean might be best, but maximum and minimum values are also important (they may be correct anyway), as is standard deviation. These are important statistics which convey to people (who weren't there) how precise your figures are likely to be. The way you convey the precision of your results will depend on your method. In interviews with local people, how many people said one thing and how many people contradicted this? During long watches, there may be doubt as to which bird was which – in this case what was the minimum and maximum numbers that you are sure of?

Section 5
ASSESSMENT OF SITES: MEASUREMENT OF SPECIES RICHNESS AND DIVERSITY

Peter A. Robertson & Durwyn Liley

5.1 Introduction

Many projects aim to assess the conservation importance of a site or the relative values of different habitats and do this by determining the diversity of species present. Such surveys provide baseline conservation data on the distribution of key species, the richness of sites or habitats and allow comparisons to be made between areas. For such data to be meaningful it is necessary to know how accurate and how complete they are.

In tropical forests it is notoriously difficult to locate birds. The very structure of the habitat, with high canopies and sometimes with dense undergrowth, means that birds are difficult to see. Many species occur at very low densities and the difficulty of identifying species compounds the problem. With a high diversity of possible species, accurately describing the avifauna of a tropical forest site presents considerable problems, particularly if time is limited and the observers are unfamiliar with the species present. For these reasons it may be excessively time consuming and unrealistic to collect density information as described in the previous section. Measurement and comparison of species richness may be a sufficiently challenging and worthwhile aim.

This section is concerned with methods by which a team can collect data which are as complete and meaningful as possible, given the constraints of the habitat, observer experience and time. Methods are discussed by which it is possible to record the bird species present at a site, to determine how complete the list compiled is, to judge how much time is needed to sample a site to some degree of accuracy and lastly, to compare between sites.

5.2 Compiling a species list

The most basic description of the avifauna of a site is a species list. A list describes the diversity of a site, and shows the presence or absence of rare species. Species that are globally threatened (Collar *et al.* 1994) are key species for conservation and as such are key species to locate on any bird survey. The number of rare species and the diversity of species present at a site can be used as indicators of the importance of different sites or habitats for bird conservation. It is important that a species list should be as exhaustive as possible or that its incompleteness should be acknowledged and understood.

Compiling a species list is principally a matter of spending time birdwatching at a site. However, while the number of species recorded is largely dependent on the time spent in the field, there are a number of techniques that can be used to maximise the variety of species recorded in a short time. They amount to the skilled use of a variety of observation techniques supplemented with the use of supporting equipment. These techniques are described below:

Habitats
The full range of habitats and altitudes at a site should be covered. Subtle habitat variations can be important. Habitat breaks and changes in habitat such as ridges and valley bottoms are good areas to focus on, as are streams and marshes, particularly in dry regions or during the dry season. Forest edges are well known to attract birds and can provide easier viewing. Many species depend on restricted habitats such as bamboo clumps and it is therefore important to locate and search such restricted habitats, particularly as the species restricted to these habitats are often of conservation concern.

Canopy watching
Many species in tropical forests are more or less restricted to the forest canopy and can be difficult to see because of the height of the canopy above the ground. By making use of high ground, slopes, knolls, hill sides or by climbing a tree it is possible to be level with the canopy and increase the chance of seeing canopy species. If such an opportunity does not present itself then an observer could choose a spot with an unrestricted view of the canopy and lie on his/her back to allow a prolonged period to be spent concentrating on the canopy whilst avoiding neck ache!

Sky watching
Vantage points providing a view over the canopy increase the likelihood of seeing raptors, swifts, parrots and other frugivores. Some species (e.g. some parrots and some pigeons) roost communally and can be counted flying to and from the roost at dawn and dusk.

Speed of walking
A fast, quiet pace is better for detecting ground birds on a path and for encountering flocks; slow walking is better for detecting species in the canopy and away from a path. Skulking understorey species can be detected by scanning ahead along paths and stretches of streams, particularly when first rounding a bend in a path or stream. Frequent stops, listening for the

movements of understorey species, such as the rustling of leaves, can also help to detect this difficult group of species. If an understorey species is flushed without being identified, waiting silently in the spot or leaving and cautiously returning shortly afterwards may allow the same individual to be seen and identified.

Flushing/rope-dragging

Rails and other shy waterbirds in marshes and larks in grasslands can easily escape detection by a single observer. Sometimes the only way to detect these species is for a group of observers to walk in a line across an area of suitable habitat in an attempt to flush any individuals present. A fast pace is required to avoid birds fleeing without breaking cover. Similarly, a rope can be dragged across the top of the vegetation in marshes and grasslands to flush birds. The rope should be thick enough to disturb the vegetation without being too heavy to drag. Nylon ropes are preferable in marshes as they do not absorb water. Densities can be calculated from rope-dragging by calculating the area of habitat disturbed and assuming that all birds present were flushed.

Sitting still

Certain points such as fruiting trees, streams, pools, breaks in undergrowth or bamboo provide good vantage points for waiting for birds to appear. Patience can be rewarding.

Timing

Activity patterns vary between species. In West African forests the frequency of calling of many species decreases after 0930h. Mist-netting studies in central America have shown that certain groups of species were active at different times of day, for example more than half of all species trapped, and most Tyrannidae, were trapped in the early afternoon. Between 0900 and 1200h is the time of peak activity for most soaring raptors. It is therefore necessary to carry out searches at different times of the day and not concentrate exclusively on the early morning period. Blake (1992) describes the effect of timing on the results of point counts in a lowland wet forest in Costa Rica.

Nocturnal species

As an extreme example of different activity peaks between species, certain birds, particularly owls and nightjars, are active only at dawn, dusk and at night. These species are often under-recorded and it is necessary to spend time in the study area at night in order to stand a chance of recording them. Many nocturnal species are very vocal and can usually be identified by call alone, although a powerful torch is also useful. A tape recording of the calls of possible species can be used to elicit a response.

Knowledge of calls

Knowledge of the calls of target species and of shy or skulking species will greatly increase the chances of recording these species at a site. Tapes of bird calls from many parts of the world are available (see appendix at the end of this chapter) and can be used to learn calls before starting fieldwork, thus saving time and energy during survey work. Unfamiliar calls heard while in the field can be tape recorded (see below) or transcribed into a notebook and identified later by reference to pre-recorded tapes.

Use of tape recorders

Small portable tape recorders and speakers are available relatively cheaply and can be of great help in the field. Playing the call or song of a species will often produce a response if there is an individual of that species within earshot of the tape recorder, with the bird either coming out into the open or calling in reply. The chance of encountering shy, skulking or quiet species and nocturnal species can be greatly increased by tape playback. Walking through suitable areas occasionally playing calls of potential species is a possible method. In addition, the use of a microphone enables an unknown call to be recorded and played back immediately to bring the bird in question out into the open. When using these techniques the welfare of the bird should always be carefully considered as the excessive use of tape playback can cause disturbance to breeding birds. Marian *et al.* (1981) give a detailed account of playback techniques and their possible side effects.

Attracting species

Some species can be attracted to a particular spot allowing observers to record their presence. Bait can be used to lure species, for example fruit for attracting both frugivores and insectivores feeding on insects attracted to the fruit, honeycomb to attract some species of honeyguide and sugar solution to attract hummingbirds and other nectar feeders. In an otherwise dry area, drinking pools can be created which attract some species to drink, particularly at the hottest part of the day. Certain noises will also attract birds to the observer; 'pishing' is a well known technique among birders, a squeaking sound made with pursed lips and often the back of the hand which can draw passerines in close. It is also possible, once learnt, to draw in flocks by imitating owl species, for example, the call of a Ferruginous Pygmy-Owl in South America or the Barred Owlet in East Africa. Alternatively, recorded calls of these species could be played on a loop tape.

Special events

Certain events within a tropical forest tend to concentrate birds from a wide area. Such events include trees fruiting or army ant swarms. Although these events are unpredictable in terms of when and where they will occur, it is worth putting some energy into locating them as they are often focal points for the activity of many species within a forest. Mixed-species groups of birds will often forage together as a flock. If such a group is encountered then it is worth following until all the species present have been identified.

Mist-netting

Mist-nets can be used to add species to a site list, but training is necessary to learn how to set the nets and more particularly how to remove birds from them. The use of mist-nets by people with no training represents a serious threat to the birds caught and should not be attempted. In some countries there is a legal requirement for those who use mist-nets to possess a licence. However, for the appropriately qualified, mist-nets are an effective means of detecting skulking understorey species. They can also be used to catch species in the canopy, although this is more difficult and capture rates tend to be much lower than in the understorey making the effort necessary much less worthwhile. Mist-nets are most frequently used to survey understorey birds, and the number of birds caught will, to some extent, depend on where and how the nets are set. Nets set by water, fruiting trees and low vegetation are likely to be effective in catching birds. Some observation of areas of activity can be useful in determining the best location to set nets. The use of mist-netting as a quantitative survey technique is dealt with later (Section 5.3.5).

Knowledge of the ecology of species and targeted searches

After the initial survey effort when the common species will have been found it can be very helpful to review the list of possible species not yet recorded, especially targets such as threatened species, and focus continuing searches on these using a knowledge of the species ecology to select the methods from the list above which are most likely to locate these species.

Local knowledge

Local people, particularly hunters and harvesters, often have a very good knowledge of many of the species present at a site. Even unstructured, informal interviews can be used to provide observers with an idea of species that could be expected before visiting the area. The absence of key species, particularly quarry or otherwise significant species, can also be concluded from interviews with the local population. Simply showing someone a field guide to an area and noting which species are recognised and which are unfamiliar is a good technique but should be used with a sensible degree of precaution, questioning to check reliability where possible. The number of species apparently recognised that are highly unlikely to occur in the region gives a good indication of accuracy. If such interviews are carried out in a number of communities around a site, the consistency with which a certain species is reported by different communities, groups or individuals can be used as a means of measuring the reliability of the record, although some errors may persist between interviews.

5.3 Standardising recording methods

Species lists can show considerable variation in how accurately they describe the avifauna of a site. It would obviously be unwise to compare two species lists from different sites if one was collected over a two week period and the other was the result of many years of data collection with repeated visits to a site by a variety of observers. Lists can vary according to factors such as the length of time over which the data was collected (with more of the normally resident species being recorded over a longer time period but also more species of only chance occurrence), the quality of the observers and the variety of habitats sampled. The value of a species list is greatly increased if the level of effort is measured and if the techniques used in compiling the list are standardised. Annotating a list with the relative abundance of each species also makes it much more useful. There are a variety of ways in which effort can be measured and relative abundances can be calculated.

5.3.1 Species discovery curves

The importance of the amount of time spent at a site has already been stressed. The frequency of adding new species to a list declines with time; at the start of fieldwork every species recorded will be new and as time spent in the field increases so fewer and fewer new species will be recorded. Yet even after spending months at some sites it is still possible to add new species to the list. When collecting data for a species list the time period over which the list was collected, the number of observers and the number of hours spent in the field should be recorded. By also noting the time and date at which each new species is recorded some simple analysis becomes possible.

The rate of species discovery can be recorded in the field by dividing the overall survey effort at a site into standard units, and recording all species noted during each unit. Survey effort is a function of the time spent surveying and the number of observers. Each observer is only collecting independent data if he is working at a different place from other observers. Thus, if observers are working in pairs (e.g. for safety reasons) each pair is effectively acting as a single observer. Units of effort should therefore be observer x unit of time (e.g. observer hours or observer days). Simultaneous periods of observation by different observers can be grouped together or treated as consecutive periods of observation (e.g. four observers working in different areas during the same one hour period can be treated as one unit of four observer hours or as four consecutive units of one observer hour). The unit of time used may vary from an hour to a day (or even longer). The advantage of using a day as the time unit is that the activity pattern of species is approximately the same for each recording unit, although this is only practical if at least ten days, and ideally a rather longer period, is to be spent at the site. If less than a day is used as the time unit, then changes in species activity patterns will affect the species discovery curves, for example with fewer new species likely to be discovered during a time period covering the middle of the day. However, a shorter time unit will give more detail in the curve, particularly in a short overall recording period and the problems of changes in daily activity patterns can be reduced by not recording over the middle of the day. The recording unit of time should be chosen such that a minimum of at least ten recording units make up the total period of observation.

If the cumulative total of species recorded is plotted against survey effort then a curve rising to a plateau will result (Figures 21 and 22), as fewer and fewer new species are discovered with continuing effort. Such a curve can be a useful indicator of the optimum length of time to spend at a site illustrating when the majority of species have been recorded. The position of the plateau of the curve can be used to compare species richness between sites. Figure 21 shows a graph from a site in Indonesia, the data for which was collected over a four month period from a forest block of $3km^2$. It is only after approximately 50 days of fieldwork that the majority of species have been recorded.

Figure 21. Species discovery curve. The graph shows the cumulative total of the number of species seen during fieldwork between August and November in West Java, Indonesia. The site included a range of primary forest, secondary forest, forest edge and scrub habitats (from the authors' own data).

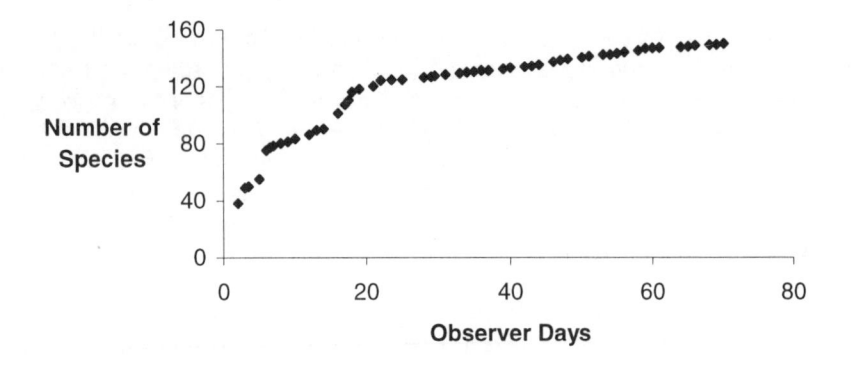

Figure 22 gives results from surveys carried out in three different forest types on the Freetown Peninsula, Sierra Leone, West Africa (Ausden and Wood, 1991). It shows a slow but continuing rise in the closed canopy forest compared with a relatively rapid rise towards a plateau in the cumulative total of species in the secondary regrowth. It also shows a shoulder in the curve for degraded forest after 70 man hours, which results from a change of base camp to a new part of the forest. This illustrates the importance of thorough coverage of the site if results are to be applied to the whole site.

Figure 22. Species/time curves for the three habitats surveyed (from Ausden and Wood, 1991)

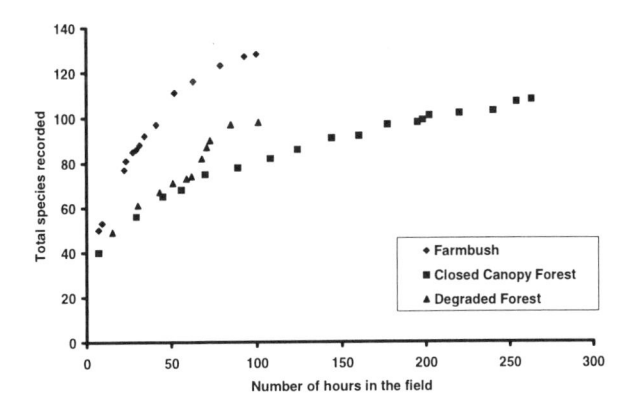

5.3.2 Encounter rates

One means of incorporating the effort expended into the analysis of bird survey results is to record field hours for each observer and the number of individuals of each species observed. This allows an encounter rate to be calculated for each species by dividing the number of birds recorded by the number of hours spent searching, giving a figure of birds per hour for each species. Additional information can be gained by determining separate encounter rates for each broad habitat type (e.g. primary forest and logged forest). At each site all the main habitat areas accessible should be visited.

The data provided by encounter rates do not provide an accurate indication of abundance and are not a substitute for density estimates. However, if it is assumed that a species is as easy to locate at one site as another then the encounter rates are crudely comparable for a species between sites. Encounter rate data can be split into crude ordinal categories of abundance (e.g. abundant, common, frequent, uncommon and rare), making these terms much more useful as they have some definition and allowing a species list to be annotated in such a way that future surveys might detect large scale changes in the abundance of individual species. An example of abundance categories related to encounter rates, used on an expedition to Paraguay, is given in Figure 23.

Figure 23. Using encounter rates to give a crude ordinal scale of abundance (from Lowen et al. 1996).

Abundance category (Number of individuals per 100 field hours)	Abundance score	Ordinal scale
<0.1	1	Rare
0.1–2.0	2	Uncommon
2.1–10.0	3	Frequent
10.1–40.0	4	Common
40.0+	5	Abundant

One important bias in the use of encounter rates in the field is that, with all the observers starting from the same base at the start of fieldwork, any species roosting near the site will be recorded by all observers at the beginning and end of each day. This bias can be reduced if observers move rapidly to a starting point some distance from the base camp, and if these starting points are varied between observers and between days. Vocal and prominent species are also over recorded at the expense of more skulking or quiet species, and this factor must be taken into account when describing the results. Although this factor will most commonly prevent comparison of different species, it might also prevent comparison of the same species at different times if there is some seasonal change in the detectability of the species, for example a difference in frequency of vocalisations between the breeding and non-breeding seasons. These factors are commonly but mistakenly ignored so that some species described as rare are actually just difficult to detect and some species described as common may just have obvious and far-carrying calls.

5.3.3 Mackinnon lists

Mackinnon Lists (Mackinnon and Phillips, 1993) provide another means of calculating a species discovery curve and an index of relative abundance. Mackinnon Lists differ from the other techniques in that the unit of effort is the time taken for an observer to record a pre-determined number of species. The advantage of this is that it makes the method relatively less susceptible to differences in ability and concentration of the observer. If an inexperienced observer takes a long time to identify each species detected this does not greatly affect the results providing he/she does eventually identify all species detected. Similarly recording during a period of low activity such as over midday will not greatly affect the results – it will just take longer to detect a given number of species.

The observer makes a list by recording each new species until a predetermined number of species is reached. A species can only be recorded once in each list but may be recorded in subsequent lists. The appropriate length of list can vary between 8 and 20 species; the larger the likely total number of species at the site the longer the length of list chosen. Comparisons can only be made between surveys where the same length of list was chosen. Surveys are repeated until a minimum of ten and preferably more than fifteen lists have been produced for each site. When recording data the observer is free to search for birds in as efficient a manner as possible, using whatever search techniques from section 5.2 are appropriate for the site. However, the observer should endeavour to cover different ground at least from one list to the next to avoid recording the same individuals on repeated lists. A species discovery curve, as described above, can then be drawn by replacing the unit of survey effort with the number of lists and plotting this against the cumulative total number of species. As above the position of the plateau of the curve reflects species richness and the shape indicates how many more species are still likely to be found in that locality (see Section 5.4.1 for analysis of Species Discovery Curves). Figure 24 shows an example of such curves from four localities in Indonesia, the shape of curve and steepness differs between sites.

Figure 24. Species curves from four different sites in Indonesia. (From Mackinnon & Phillips, 1993. Reproduced with permission of Oxford University Press)

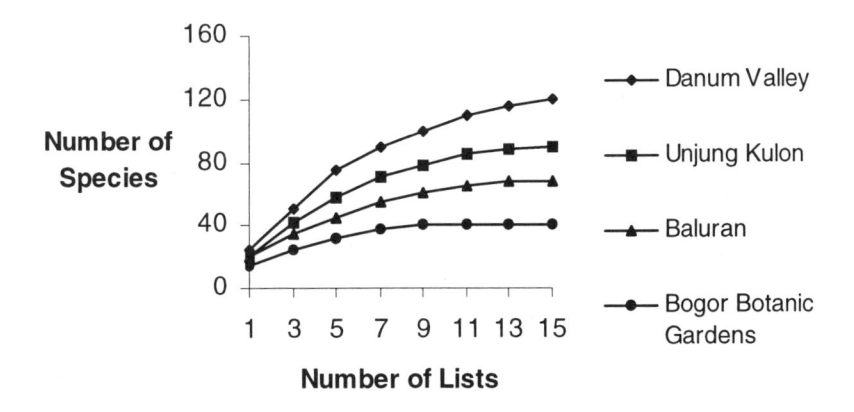

5.3.4 Timed species-counts (TSCs)

Timed Species-Counts (TSCs) were developed by Pomeroy and Tengecho (1986) for open woodland and bush habitats. They provide a simple method of comparing the avifaunas of extensive areas by sampling representative habitats. Simple, rapid and effective, they give a reasonable measure of relative abundance. They are best suited to extensive areas of open habitats and there are limitations to using the method in thick forest. Bennun and Waiyaki (1993) give an example of the use of TSCs in Kenya.

Data for TSCs are recorded in six columns, corresponding to six 10-minute intervals during an hour long survey. The observer walks at a slow pace (approximately 1–2 km/h^{-1}) through the study area for one hour. For the first ten minutes, every species recorded is noted down in the first column, giving only the species name, not the number of individuals. For the second 10-minute period, any species not already recorded is noted in the second column. The remainder of the hour is also divided into 10-minute periods and any species recorded for the first time during any 10-minute period is noted down in the relevant column, such that every species recorded during the hour is written down only once, in the column relating to the 10-minute period during which it was first seen. The analysis (described in Section 5.4.4) gives an index of relative abundance based on the assumptions that the more common species will be recorded earlier during each survey and in more different surveys than rarer species.

A minimum of 15 surveys should be carried out at different parts of the site. The choice for the length of time for each count is a trade off between recording as many species as possible and not spending so long that few visits can be fitted in. An hour of observation is recommended as standard so that surveys by one team can be compared with similar surveys at different sites or by different teams at the same site. Pomeroy and Tengecho (1986) also recommend using an area of 1km^2 for each count. In making the count the observer intentionally visits as many parts of the area as possible and concentrates on areas where bird activity is greatest.

Simple habitat and environmental variables can be included in the analysis which will help to account for differences in bird communities. Pomeroy and Tengecho recorded two variables: moisture index and percentage cover of woody plants. Other variables could also be used (see section 6).

5.3.5 Mist-netting

Mist-netting is an effective means of recording quiet and skulking species of the forest understorey which may not be recorded using other techniques.

Standardised mist-netting effort can be used to compare this element of the avifauna between sites. However, mist-netting is a very labour intensive technique and the purpose of using mist-nets should be carefully considered before any netting is undertaken. They should only be used by people with appropriate training to avoid any threat to the birds caught.

For comparison of sites a minimum of 100m of nets set in a standard form (e.g. on a random grid or in a straight line) should be used with at least two days of catching effort. Nets should be opened for 3–4 hours from dawn and perhaps also for two hours before dusk each day, although capture rates in the evening are often lower than those in the morning. The results from standardised mist-netting studies are expressed in units of birds caught per metre per hour (i.e. birds $m^{-1}h^{-1}$). Marking individuals caught (with metal rings or colour rings) allows re-captured birds to be excluded from capture totals. This technique should only be used by those with appropriate training. Nets should be checked at least every half hour or more frequently if it is very hot, and should be closed immediately during rain. The effectiveness of mist-netting is greatly affected by the skill with which the nets are set and also by the type and condition of the nets. More birds tend to be caught if the nets are set along lines specially cut through moderately dense vegetation than if existing tracks or open areas with little understorey vegetation are used. Nets set parallel or perpendicular to streams or close to other water sources tend to be particularly effective but such selective positioning of nets reduces the comparability of results between different sites.

5.4 Analysis of data

There are a number of ways in which the data collected from the techniques described above can be analysed. The purpose of such analysis is to enable comparisons to be made between sites and allow predictions to be made of total species richness of sites, to describe the relative abundance of different species within a site and the relative abundance of the same species between two or more sites. Rigorous recording of the methodology used and particularly the effort expended is essential to allow such comparisons.

5.4.1 Predicting total number of species from species discovery curves

Two types of species discovery curves have been described, using observer time units as a measure of survey effort (Sections 5.3.2 and 5.3.4) and using repeated species lists as a measure of survey effort (Section 5.3.3). In each case, a curve can be fitted to the points plotted. If the curve is extrapolated beyond the data obtained, a prediction of the level at which it reaches a plateau can be made. The level of this plateau is equivalent to the total number of species expected at the site. Plotting such a curve also enables an estimate of the level of effort required to add a given number or percentage of species, allowing approximate calculations to be made of the amount of time to spend at a site in order to optimise the number of sites to be visited in a given time. This analysis is best carried out on a computer with a suitable statistical package, although the choice of model used to describe the curve is not simple. Two alternative general curve shapes are a logarithmic curve and an exponential curve. An exponential curve may be adequate when a well known avifauna is sampled in a small, homogeneous area, but where a poorly known avifauna is sampled in a large and heterogeneous area the logarithmic curve may be preferable. Trying out different curves with a computer package allows the curve with the lowest residual variance to be selected. Alternatively the curve could be plotted by eye.

An alternative method of predicting the species richness of a site is to plot the number of new species recorded for each unit of survey effort against the log of the cumulative number of species recorded prior to that unit of effort: a linear relationship is expected. The number of additional species will decrease inversely with the cumulative total and a regression line can be plotted using a computer statistics package. Where the line of best fit crosses the x axis an estimate can be obtained of the total number of species present. An example from Kenya is shown in Figure 25. Because a straight line relationship is expected it is easier to fit the line by eye than it is to fit a curve as described above.

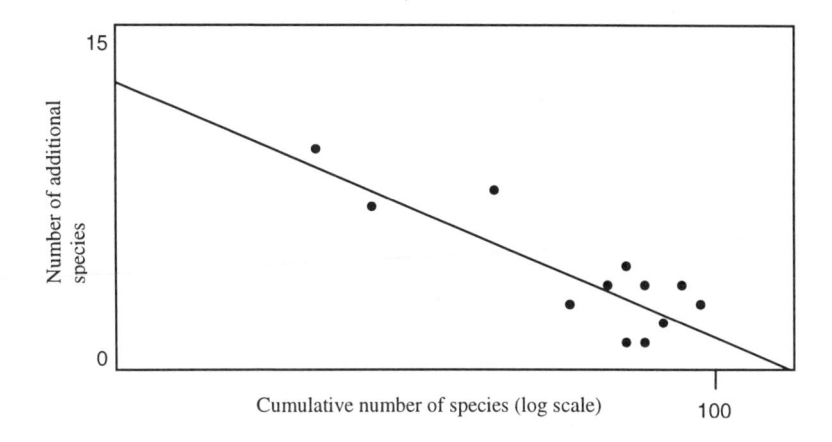

[Graph: Number of additional species vs. Cumulative number of species (log scale), with data points and a regression line crossing the x-axis near 100.]

Figure 25. An example of analysis used to predict the total number of species at a site using cumulative count data (from Pomeroy and Tengecho, 1986). The point at which the line crosses the x axis gives an estimate for the total number of species present at the site and is 120. $R^2 = 0.880$, $P<0.001$, for the regression line sa = 34.8–7.3 (log_{10} sp) where sa is the number of species at successive counts of 1.0 hours and sp is the cumulative number of species.

5.4.2 Analysis of encounter rate data

The analysis of encounter rate data is very straightforward. Figure 26 shows the results from a survey carried out by three independent observers in a forest in Madagascar. Observer 1 spent two hours in the forest while observers 2 and 3 spent three hours each. The total observation period is thus eight hours. The encounter rate for each species is equal to the total number of individuals recorded by all three observers divided by the period of observation and multiplied by ten to give a result in units of number of individuals recorded per ten hours of survey.

Figure 26. Encounter rates from a forest in Madagascar (author's own data). There were eight survey hours for each observation.

Species	Number of individuals by each observer (3 observers)			Number of individuals /10 hours	Class or relative abundance
	1	2	3		
Frances's Sparrowhawk	1			1.25	Uncommon
Madagascar Buzzard	1		1	2.50	Frequent
Grey Cuckoo-shrike	2	2	1	6.25	Frequent
Madagascar Blue Pigeon	4	3	2	11.25	Common
Madagascar Turtle Dove		2	1	3.75	Frequent
Madagascar Black Coucal		1		1.25	Uncommon
Blue Coua	4		4	10.00	Frequent
Giant Coua		1		1.25	Uncommon
Red-fronted Coua	1	2	2	6.25	Frequent
Mascarene Martin	1			1.25	Uncommon
Cuckoo Roller	2	2	1	6.25	Frequent
Madagascar Sunbird			2	2.50	Frequent
Souimanga Sunbird	11	8	9	35.00	Common
Red-tailed Vanga	2		1	3.75	Frequent
Blue Vanga			2	2.50	Frequent
Madagascar Drongo	4	7	2	16.25	Common
Chabert's Vanga		2		2.50	Frequent
White-headed Vanga			1	1.25	Uncommon
Hook-billed Vanga	1	1	2	5.00	Frequent
Velvet Asity	1			1.25	Uncommon
Madagascar Fody	1	1		2.50	Frequent
Forest Fody		8		10.00	Frequent
Grey-headed Lovebird		7	4	13.75	Common
Lesser Vasa Parrot	2	1	7	12.50	Common
Madagascar Magpie-Robin	1	2	1	5.00	Frequent
Madagascar Bulbul	29	12	17	72.50	Abundant
Common Jeery	4	8	2	17.50	Common
Madagascar Scrub Warbler	8	4	6	22.50	Common
Common Newtonia	2	6	5	16.25	Common
Madagascar White-eye	6	8	8	27.50	Common

5.4.3. Analysis of Mackinnon list data

The data gathered by Mackinnon lists can be analysed in two ways. Species discovery curves can be produced by plotting the cumulative total of species against the number of lists and analysing the results as described in section 5.4.1. In addition, the results can be analysed to give an index of relative abundance for each species. The relative abundance of each species at each site is equivalent to the fraction of lists on which a species occurs, i.e. if a species is recorded on 8 out of 10 lists at site A and on 3 out of 15 lists at site B then its index of relative abundance is 0.8 at site A and 0.2 at site B. This index can vary between 0 (species not recorded) and 1 (species recorded on every list).

5.4.4 Analysis of TSC data

In analysing the results (Figure 28, in which survey 1 is that shown in Figure 27), each species is given a score depending on the 10-minute period in which it was first recorded, such that species recorded in the first ten minutes are given a score of six, species first recorded in the second ten minutes given a score of five and so on, with species recorded in the final ten minutes being given a score of one. If a species is not recorded from a survey then it has a score of zero for that survey. An index of the relative abundance of species recorded from repeated surveys is calculated as the mean of scores from each survey, and therefore varies between a maximum value of six and a minimum value of $1/n$ (where n is the number of repeated surveys).

Figure 27. Results from TSC in savannah forest, Sierra Leone, West Africa (from authors' own data).

0–10 mins.	10–20 mins.	20–30 mins.	30–40 mins.	40–50 mins.	50–60 mins.
Common Bulbul; Great Blue Turaco; African Golden Oriole; Grey-headed Sparrow; Black Kite; Tawny-flanked Prinia	Turati's Boubou; Red-chested Cuckoo; Bush Petronia; Blue-spotted Wood Dove	Dybowski's Twinspot; Red-eyed Dove; Swallow-tailed Bee-eater	Black Cuckoo; Yellow-winged Pytilia	Vieillot's Barbet; Red-winged Warbler	Grasshopper Buzzard; Yellow-throated Leaflove

Figure 28. Analysis of TSC results from four surveys.

Species	Survey scores				Total score	Mean score	Species rank
	1	2	3	4			
Common Bulbul	6	6	6	6	24	6	1
Great Blue Turaco	6	4	4	2	16	4	7
African Golden Oriole	6	5	5	3	19	4.75	5
Grey-headed Sparrow	6	4	5	6	21	5.25	3
Black Kite	6	6	2	3	17	4.25	6
Tawny-flanked Prinia	6	6	5	5	22	5.5	2
Turati's Boubou	5	5	4	6	20	5	4
Red-chested Cuckoo	5	3	2	4	14	3.5	9
Bush Petronia	5	3	2	2	12	3	10=
Blue-spotted Wood Dove	5		3	3	11	2.75	12
Dybowski's Twinspot	4	2		2	8	2	15
Red-eyed Dove	4	4	4	3	15	3.75	8
Swallow-tailed Bee-eater	4			3	7	1.75	16
Black Cuckoo	3		4	2	9	2.25	13=
Yellow-winged Pytilia	3	3		3	9	2.25	13=
Vieillot's Barbet	2	5	3	2	12	3	10=
Red-winged Warbler	2	3	1		6	1.5	17
Grasshopper Buzzard	1	1			2	0.5	18
Yellow-throated Leaflove	1				1	0.25	19

5.4.5 Analysis of mist-net data

A survey comparing the communities of different forest types around Gola Forest, Sierra Leone, West Africa (Allport *et al.* 1988) used mist-netting as one of a variety of methods. The results were analysed using diversity, equability and similarity indices. These are a convenient, although not foolproof, means of combining the species richness (total number of species) and the evenness (extent to which all species are equally common) of a bird community and as such may be used whenever there are data available on number of species and their relative abundance. The most widely used indices of diversity and equability are the Shannon–Wiener indices;

Diversity Index $H' = - \Sigma p_i . \ln (p_i)$

where p_i is the proportion of species *i* expressed as a proportion of the total number of individuals of all species, ln is the natural logarithm, and Σ represents the total $p_i . \ln (p_i)$ for all species.

Equability Index $J = H'/H_{max} = \Sigma p_i . \ln (p_i)/\ln (s)$

where s = number of species.

The results from Gola Forest are shown in Figure 29.

	Total Catch		43 bird sub-sample			
Forest Type	Diversity	Equability	Diversity	Equability	Number of birds	Number of species
Secondary Regrowth	2.72	0.88	2.58	0.88	56	22
Plantation	2.31	0.81	2.13	0.81	50	17
Logged	2.75	0.92	2.58	0.91	65	20
Primary	2.59	0.91	2.59	0.91	43	17

Figure 29. Analysis of mist-net catches in different habitats around Gola Forest, Sierra Leone, W. Africa (author's own data).

The results were analysed for the total catch in each forest type and for a sub-sample of 43 birds, the minimum number caught in any one forest type. The higher the diversity index the greater the number of species and evenness of their populations. This can result in communities with higher species richness and lower evenness having the same diversity index as communitics with a lower species richness and a higher evenness. Equability Indices vary from 0 to 1, with communities where all species are equally abundant having an index of 1. In the Gola example, the plantation forest understorey community has a lower diversity and equability index than the other forest types, which reflects the fact that the understorey vegetation in plantations is very uniform and 46% of the birds caught in it were of just two species. The relatively high diversity of the secondary regrowth may have been due to the fact that the mist-nets covered almost the entire height range of this habitat, so that species which in other forest types would not have been caught in the nets were caught in this forest type.

Similarity Indices measure the degree to which the species and their relative abundances are shared between different bird communities. Completely similar communities have an index of 1 while completely dissimilar communities have an index of 0. The Czekanowski similarity index is widely used;

Similarity Index $S_c = [2 \Sigma \min (X_i, Y_i)]/[\Sigma X_i + \Sigma Y_i]$

where X_i and Y_i are the abundance of species i in habitat X and Habitat Y and $\Sigma \min (X_i, Y_i)$ is the sum of the lowest abundance where species i occurs in both habitat X and habitat Y.

A similarity matrix of the four forest types sampled using mist-netting in Gola Forest is given in Figure 30.

	Plantation	Logged	Primary
Secondary Regrowth	0.991	0.516	0.475
Plantation		0.323	0.315
Logged			0.989

Figure 30. Similarity indices of bird communities from different habitats around Gola Forest, Sierra Leone, W. Africa (from mist-net data) (author's own data).

This analysis shows the similarity between the secondary regrowth and the plantation and between the logged and primary understorey forest communities, and the lack of similarity between these two groups.

5.5 Discussion

The methods described in section 5.2 are aimed at increasing the efficiency of compilation of a species list, i.e. ensuring that all species present are found in the shortest possible time. The methods and analyses described in sections 5.3 and 5.4 are aimed at standardising the application of the methods described in 5.2 to allow comparisons between sites and between surveys. The usefulness of a species list is significantly increased if the methods by which it is compiled are carefully designed. It is necessary to include or describe observer experience, effort, habitat, time of day and season. Species Discovery Curves are a means of estimating how much of the total bird community has been recorded during a survey and the total number of species that is likely to be recorded. Encounter rates, Mackinnon lists and Timed Species-Counts are simple methods of making an estimate of the relative abundance of species and may be used to compare different species within a site or the same species between different sites. However, such comparisons must be made with care, taking into account the possible effects of various biases on the results. Perhaps the most significant bias in the use of methods measuring the relative abundance of species is the difference in detectability between species. Thus, rare but large and vocal species may be recorded as more common than common but small and secretive species. The effect of this bias can be reduced for all the methods by setting a distance limit of say 20m on all records, such that any individual recorded at a distance of greater than 20m is discounted. This will, however, greatly reduce the amount of data gathered. Alternatively data could be gathered from two width bands of less than and greater than 20m and the results analysed separately or combined. A more effective way to rule out this bias is to use methods based on distance estimation (Section 3). However, the rigorous methodology of these methods requires preparation before the survey can start, e.g. cutting transects or marking random points; limits the observer's freedom to apply the search methods described in section 5.2; and require the gathering of a relatively larger amount of data before a valid analysis, which is in itself complex, can be carried out.

The quality of data gathered by Mackinnon Lists is relatively unaffected by the skill of different observers and by the change in expertise of a team as it gets to know a bird community. If an inexperienced observer takes a long time to identify each species seen he/she may still compile the same list as an experienced observer although it would take him/her longer to do so. This would not effect the results as this method is not based on units of time as is the case for encounter rates and Timed Species-Counts. Similarly, lapses of concentration, which are likely to occur during long periods of observation, would have relatively little effect on the results.

Mackinnon Lists and Timed Species-Counts tend to underestimate the abundance of species which are found in flocks, as they take no account of the number of individuals encountered. Thus, a flock of thousands would be recorded the same as a single individual using Mackinnon lists and Timed Species-Counts. The difference between a flock and an individual is only taken into account when using encounter rates.

Mist-netting can be used to record species in the understorey which might otherwise go unrecorded and can also be used to estimate the relative abundance of understorey species.

Figure 31. Summary of uses, advantages and disadvantages of the methods described in this section

Method	Uses	Advantages	Disadvantages
Simple Species List	Determining which species are present.	No analysis involved. Simple.	No account taken of effort, thus difficult to make comparisons between surveys and between sites.
Species Discovery Curve	Analysing the completeness of a species list, estimating the total number of species present and comparing lists from different sites.	Allows comparisons between species lists for the same site over time or for different sites.	Accurate plotting of curves requires computer analysis.

Encounter Rate	Annotating a species lists with an index of relative abundance based on the number of encounters with individuals per block of time.	Allows crude comparisons of abundance between species within a site and within species between sites.	Prone to bias resulting in differences in species' detectability. The requirement to count all individuals of all species presents practical problems.
Mackinnon Lists	Annotating a species list with an index of relative abundance based on the number of encounters with species per block of effort. Plotting a species discovery curve.	Allows crude comparisons of abundance between species within a site and within species between sites. Data collection is simple, allowing the observer freedom to roam. Relatively unaffected by observer skill and concentration.	Prone to bias resulting in differences in species' detectability. Under-estimation of flocking species.
Timed Species-Counts	Annotating a species list with an index of relative abundance based on the number of encounters with species per weighted block of time. Plotting a species discovery curve.	Allows crude comparisons of abundance between species within a site and within species between sites. Data collection is fairly simple, allowing the observer freedom to roam.	Underestimation of flocking species.
Mist-netting	Discovery of secretive understorey species. Index of relative abundance based on the number of encounters with individuals per unit of effort.	Rigorous.	Time consuming, requires specialised equipment and highly trained personnel.

Species lists and indices of relative abundance can be used for priority setting for conservation. It is not the number of species present which is of prime importance but which particular species are present, with judgements usually being made on the basis of the conservation status of those species present. BirdLife International's Important Bird Areas programme uses four categories for selecting priority sites:

- the presence of one or more threatened species;
- the presence of a group of species with a restricted range belonging to an Endemic Bird Area;
- the presence of a group of species restricted to a biome;
- the presence of a large number of individuals.

The methods described in this Section can be used to select priority sites for conservation by making comparisons between different candidate sites or judging an individual site against a standardised set of criteria such as those used in the Important Bird Areas programme. Choices between sites with similar species lists can be made based on the relative abundance of key species at the different sites. The data gathered by these methods can also be used as the basis of a monitoring programme, allowing comparisons within sites over time.

5.6 Sources of information for the recording of bird sounds

The Wildlife Section of the British Library National Sound Archive (NSA) has over 100,000 recordings of animal sounds. It covers over 7,500 species of birds, 770 mammals, 700 amphibians, 700 invertebrates and a few reptiles and fish. Any recording may be listened to free of charge, by appointment, at the British Library building in London. Several visitors listening to different recordings may be accommodated at the same time. Copies of recordings and spectrograms can be made to order. Advice on recording equipment and techniques can be given, including equipment loans to approved expeditions, and the NSA runs an annual workshop in conjunction with the Royal Geographical Society/Birdlife International.

Contact:
Richard Ranft   Tel +44 (0)171 412 7402/3
Curator   Fax +44 (0)171 412 7441
NSA Wildlife Section   Email: nsa-wildsound@bl.uk
The British Library   http://www.bl.uk/collections/sound-archive/
National Sound Archive
96 Euston Road
London NWI 2DB
UK

The Library of Natural Sounds (LNS) at the Cornell Laboratory of Ornithology contains approximately 130,000 recordings covering over 6,000 species of birds. Other holdings include insects, amphibians and mammals. LNS offers collaborators state-of-the-art archival facilities, database software, technical advice and field recording equipment loans:

Contact:

Gregory F. Budney
Curator
Library of Natural Sounds
Cornell Laboratory of Ornithology
159 Sapsucker Woods Road
Ithaca, New York 14850
UNITED STATES

Tel +1 (607) 254 2404
Fax +1 (607) 254 2439 or 2415
Email: libnatsounds@cornell.edu
http://www.birds.cornell.edu

Wildsounds is a supplier of recording equipment and commercially available bird recordings. Their web-site has a facility to search all available recordings by species.

Wildsounds
Cross Street
Salthouse
Norfolk
NR25 7XH
UK

Tel/Fax +44 (0)1263 741100
Email: wildsounds@poboxes.com
http://www.wildsounds.com

Section 6
BIRD-HABITAT STUDIES

Colin Bibby, Stuart Marsden and Alan Fielding

6.1 Why study habitats?

Understanding the habitat associations or usage by a species is fundamental to getting to know about its conservation status. Such information can be collected by direct or indirect field methods. Beyond the level of descriptive natural history, there will be a need for some mathematical treatment of data to identify important habitat associations. As emphasised throughout this book, analysis of data depends very much on having good data in the first place and this depends on having methods well suited to the purpose. Habitat data will usually be collected at the same time as the bird data.

Ecology is the study of relationships between organisms and their environment, where the environment consists of a set of gradients or categories. Simple gradients such as altitude are relatively easy to appreciate and study. More complex gradients, for example forest disturbance by humans, are more difficult to characterise and often may only be described mathematically. Any one place, or the requirements of a bird, may lie on several independent habitat gradients. Explaining the patterns of relationship between birds and habitats is the challenge for the researcher. The value of such an approach may come from:

- predicting distribution and numbers in unsurveyed areas;
- providing an understanding of the nature of the relationship between a bird and its habitat;
- predicting possible consequences of future changes of land use.

It may be necessary to adopt a variety of methods to meet all three of these aims. Different kinds of investigation take place at different scales, and this will influence the kinds of field measuring that are possible.

We have separated the large and fine scales. Large scale approaches lend themselves to questions about distribution and numbers and extrapolation to areas not actually surveyed. Their planning and measurement relate to the design stage of studies. Fine scale studies focus on individual birds and address questions about habitat relationships in more detail. Predicting the consequences of land use changes may require data at both scales.

6.2 Broad scale habitat studies

The great advantage of a broad scale study is that you can be more confident about the area to which your findings refer. The downside is the large amount of preparatory work required and the demand for access to a bigger area. The approach requires mapping the habitats of the whole target area, which might measure up to thousands of square kilometres.

6.2.1 What features to map?

Deciding what to map is a trade-off between the kinds of features that are possible to map and those that are known or suspected to be significant for predicting the occurrence of birds. What is possible is influenced by the size of area to be mapped and by the availability of data.

Altitude

If you were only allowed one factor to map, then for many birds, altitude would have to be the choice. Wherever there is a strong topography, altitude is a major predictor of the occurrence of individual species. Altitude data are relatively easy to find from various sources as maps or digital data.

Forest structure

Numerous studies have shown the over riding impact of structure of vegetation on the distribution of bird species. For large scale mapping, the difficulty is identifying parameters of the kind which influence birds but which can also be mapped. Satellite images might be able to highlight differences between seasonal and evergreen forest. Aerial photographs might be used to separate different degrees of canopy closure. Either approach or the use of forestry maps might be able to detect the effects of logging, but this is notoriously variable in its intensity and often difficult to detect following regrowth.

Other vegetation

Structural features are also important for birds in more open habitats. It might be possible to map grasslands, scrub or savannah in terms of classes of vegetation cover. Water features are important for many birds and can readily be mapped, though there can be great seasonal variation. Satellite images are particularly good at detecting water cover. Associations with particular habitat features, such as one species of fruiting tree, or large hole-bearing trees in general, may only become obvious from more detailed studies. Generally, such features are not easy to map on a large scale.

Soils
Soil maps are readily available. Given the importance of soil types in influencing vegetation, an impact on birds might be expected. Contrary to expectation, soils do not seem to be an important factor for bird habitats. In spite of the ease of obtaining the data, this is probably not worthwhile unless you have good reason to believe otherwise in a particular study.

Human factors
The ubiquity of human influence is the constant refrain of conservation. Humans can influence birds and their habitats directly by modifying vegetation or by hunting. They can also have indirect impacts since habitat change can alter the impact of predators or allows the spread of invasive species. Indirect measures such as distance from road or village might be used as surrogate measures of human impact, so it might be worth mapping roads and villages from aerial photos, or other maps. Local people might be able to map the areas they regularly visit.

6.2.2 Sources of data for mapping
There are three common sources for producing habitat maps. They differ considerably in their costs and effectiveness.

Satellite images
Satellite images are the 'top of the range' approach but they are still expensive to purchase and process. There is a huge literature on interpretation, which is a specialist field in its own right. Provided there are no problems with shade or clouds, satellite images are good for drawing broad brush habitat maps that show major differences, such as between forest and grassland. If the images are classified without field data (the normal situation), they may map all sorts of variation, (e.g. within forest types) but these will not necessarily be clearly explicable or biologically meaningful. Maps derived from satellite images need especially careful field verification. If the field effort is sufficient, it may be possible to use the field data to produce alternative classifications of the imagery which bear more relation to the birds and their habitats.

Because they are handled on Geographical Information Systems (GIS), satellite images lend themselves to studies requiring this kind of analysis. For instance, it is very easy to measure indices of fragmentation, patch size and isolation from a satellite image.

It is possible and much cheaper to obtain satellite data as false colour printed photographs. Obviously, these cannot be used for any of the sophisticated GIS applications. They can however be used to draw generalised habitat maps to show major boundaries, such as between evergreen and deciduous forest (if taken at the appropriate time of year).

Aerial photographs
Aerial photographs are generally easier to interpret at a broad level especially if viewed as stereo pairs. They present much smaller areas at higher resolution. For this reason they are quite good for mapping plots at the lower range of sizes no more than hundreds of square kilometres. Aerial photographs are easier to find and access than satellite images. Obviously, they do not allow such sophisticated analysis. Typically you might expect to generate a map which traces up to half a dozen habitat types.

Pre-existing maps
You might be able to find previously published maps from all manner of possible sources. They may also have political significance, if for instance they were generated by a forestry service that has management authority. They have the disadvantage that they may or may not be suitable for your needs in terms of accurate discrimination of relevant habitat boundaries. They may also be substantially out of date.

Altitude as already noted is a factor that is important as a determinant of bird distribution and is reliably available from maps. Contours can also be extracted from computerised digital elevation models if you are going to handle any other data in a GIS format.

6.2.3 Verification of a map and sampling
Issues concerned with design and sampling are covered further in Section 2. Having generated a habitat map, you should use it in the design process. It should be clear from the above that some of the ways of generating the maps can be ambiguous. So you will want to verify the map at the same time as pursuing more detailed habitat assessments (described below). Essentially, you want to be able to show that the mapped units are internally consistent and distinct one from another across the range of the map. Provided you have a range of plots or samples where you have measured bird and habitat data, then you can make such an assessment in retrospect.

6.3. Fine scale bird-habitat studies

6.3.1 Different approaches/survey designs
The way you link your bird data to your habitat data will depend on the bird census method you have chosen. After all, it will be easier to collect habitat data at the same time as your bird data, rather than having to organise a separate habitat survey. It is easier to relate habitat to birds with some bird

census methods, such as point counts or line transects, than with others. Figure 32 shows some basic approaches to bird-habitat studies based mainly on the census methods described in Section 4.

Figure 32. Relating birds to habitat data using different census methods.

Census Method	Approach
Point counts	Collect habitat data at, and around the points themselves – bird presence/absence at points is related directly to the habitat there
Line transects	Transect is divided into short sections (50–100m in length) – bird presence along each section can be related to habitat directly.
Bird based studies	If focal studies are made of, say the behaviour or home range of individual birds, then the bird's use of the habitat around it is of primary importance. For example, within its home range does it stick to forest gaps, edges, or areas with dead trees? What proportion of time does it spend in each habitat 'type'? How does the habitat it uses differ from that which it does not?
Aggregation counts	Habitat recordings are made at each aggregation. The same readings might also be made at random points or places which look as if they might be suitable as aggregation sites. What features are shown by the aggregation sites and not by the areas that do not hold aggregations?
Look down methods	More difficult. Very broad habitat features (e.g. density of emergent or dead trees) can be estimated from the vantage point. Habitat readings could be taken along random transects through the survey area. Bird usage, either in different parts of a single area or between different sites could be related to the habitat features using regression analyses.

6.3.2 What habitat features to record and how?

A species' relationship to its habitat is bound to be complex. It is unlikely during a relatively short study that you will discover exactly what makes the species tick – what it needs for feeding throughout the year or for successful raising of its young. What you can do, however, is to characterise the sorts of habitats in which it seems to thrive and those from which it is absent.

The choice of habitat features to record is difficult, and will depend on the amount of time you have, the type of bird to be studied, its broad habitat and the threats that it might face. You cannot record everything and you won't want to because habitat recording is time consuming.

Altitude, gradient and other physical features

Of all possible bird-habitat associations, altitude or elevation is the most commonly cited. Details of a species' altitudinal range (maximum and minimum), the altitude at which it is most abundant, and interactions between altitude and the patterns of human habitat alteration can be crucial in defining species ranges, explaining differences in abundance between areas and hence positioning protected areas correctly. Altitude is easily measured using an altimeter or can be taken from maps if these are accurate enough.

Gradient is used less frequently but it can often be important in defining species distributions. Just as altitude has an effect on vegetation, so does gradient, with, for example, trees on steep slopes being in general smaller and those on ridges more likely to be deciduous. Birds may react to these differences, and there may be a tendency for species diversity and abundance to be lower on slopes. Gradients can be expressed as degrees or percentages and be measured or estimated at a single point or expressed as a mean gradient over several points (you could take gradient measures at four random points around each census site).

Many other physical features can be recorded, although which ones are important will depend on the type of bird/habitat you are working with. Distance to the nearest river or stream can be important in many species, while in others, the presence or absence of standing water (e.g. ponds, swamps) can be important. The presence of bare ground, rocky outcrops or boulders may be significant for some bird species. Although not strictly physical features, the distance from the census site to the forest edge, to a logging road, or to the nearest village may be potentially important. If the census site is outside a forest block, how far is it from the forest edge?

Tree sizes, tree densities and biomass indices

Tree heights and diameters at breast height (DBH) are important habitat features, which can be estimated, or measured with a tape measure at census plots, along transects, or at random points within areas. With points, a useful method is to estimate the heights (or 'measure' using an inclinometer) and measure the girths of ten trees larger than a given DBH (e.g. >0.2m DBH) which are closest to the centre of the plot. These values can be used to calculate the mean tree height and DBH at each census plot. A measure of the distance to the furthest of these ten trees will allow you to estimate the density of trees around the point using;

$$D = 100{,}000 \cdot \Pi \cdot (\text{dmax}^2) \qquad [\text{NB. } \Pi = \text{pi}]$$

where
D = tree density per hectare
dmax = distance to furthest of the 10 trees (m)

Another useful statistic is basal area of trees per hectare. This can be calculated again using distance to the furthest tree (dmax) along with the summed basal areas of each of the ten nearest trees. This measure can be extended to produce a woody biomass index, which approximates to the basal area of each tree multiplied by their heights, and can be expressed as volume of wood per hectare.

Measuring trees is time consuming and hard work, but tree sizes and densities can be very useful in describing the distributions of all sorts of birds. As well as considering trees greater than a certain DBH, it may also be worth counting saplings within a given area, the number of fallen trees, or the density of stumps left after recent timber removal. If you really do not have time to measure several trees around your census site, then measuring or estimating the girth of the largest tree, or largest two trees within your census area (e.g. within 20m of the centre of your plot) is a useful surrogate.

Tree architectures

As well as their sizes, and whether they are evergreen or deciduous, the shapes of trees can often be important predictors of bird distribution. One regime to relate tree 'architectures' to bird distributions was developed from work by Torquebiau (1986) and by Jones *et al.* (1995). Each of the ten trees nearest the census plot's central point was allocated to one of the following groups (Figure 33);

- Branching above half its height (A): trees which have grown up under the closed canopy of primary forest tend to have their first major branch well above half their height;
- Branching below half its height (B): trees which have grown up in open canopied areas usually branch below half their height;
- Branching above but with scars (C): trees which have major scars from dropped branches tend to be characteristic of regenerating forest i.e. they have grown up under a canopy that is closing up following a tree fall or logging extraction;
- Vertical branching below half its height (D): an alternative to (C) where the lower branches are maintained but grow vertically into the closing canopy.

Quantification of the above architectures, expressed as the proportions of trees at a site which show (A), (B), or (C+D) are useful indicators of the recent history of forest at census sites. Another useful measurement is the number of dead trees and stumps at sites. Relating bird presence or abundance to the proportions of different tree architectures at sites can yield information on whether the species prefers primary forest (A), recently disturbed forest (B), or older regenerating forest (C+D).

Figure 33. Tree architectures. A = branching above, B = branching below, C = branching above with scars below, D = branching below with vertical growth.

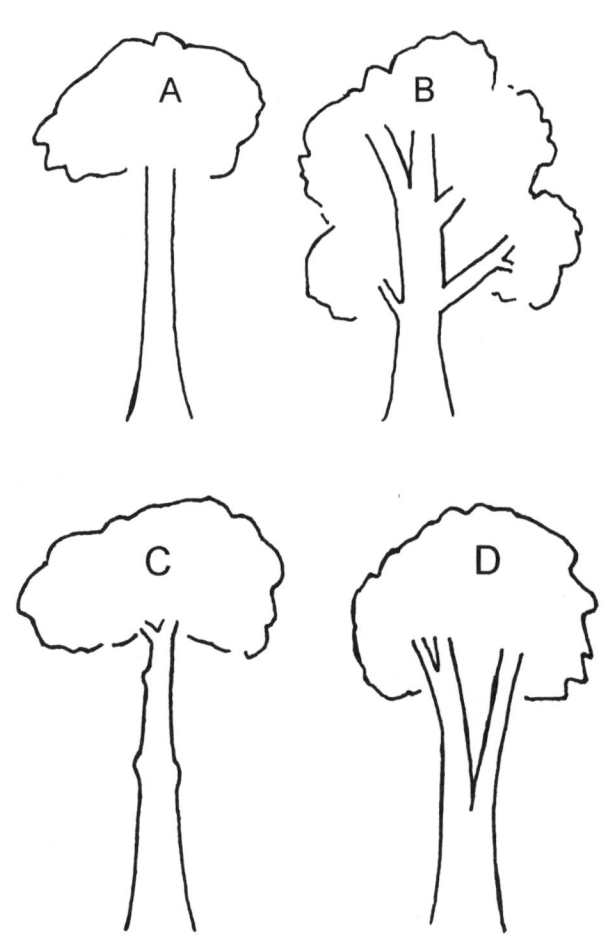

Foliage profiles/canopy covers

The percentage of vegetation cover at various strata of different habitats can, if estimated reasonably accurately, be useful in describing bird distributions. A useful regime is to estimate vegetation covers at ground level (0–1m), low-level (1–5m), mid-level (5–20?m), sub-canopy, and canopy, within a circle around the recorder of radius about ten metres. Quantification of vegetation covers is very difficult, so estimation is necessary. This should be well practised and standardised between different recorders. Disturbed forests and non-forest areas will have, in general, sparser canopy cover than primary forest, and there may be a more extensive ground and shrub or understorey layer. There can, in many habitats, be a negative correlation between the amount of vegetation in the upper and lower strata e.g. in primary forests most foliage is in the canopy and sub-canopy, while the understorey is sparse.

Plant/tree species recording

While it is desirable to relate bird distribution directly to the tree and plant species which occur in an area, this is often difficult. Tropical tree species can be very difficult to identify and most occur at extremely low densities, which makes linking a bird's occurrence directly to the presence of a particular tree species problematic. This situation may be helped if trees are lumped into genera or groups, or if a technique such as Correspondence Analysis is used to identify gradients along which 'communities' of tree species may occur.

There are some situations where recording tree species may be very useful in fairly short studies. The abundance of fruiting trees, such as figs (*Ficus*) may be related to the abundance of frugivores. Some species, such as parrots or hornbills, may nest in just a few characteristic tree species and the density of these could be recorded, and related to bird abundance. Some pioneer trees such as *Macaranga* or *Lecropia* can be characteristic of forest disturbance and the abundance of these can be used both to classify forest types and to relate to bird species occurrence or diversity.

Indices of human impact

In conservation studies, it is often important to quantify human impact on habitats. To a large extent, human impact through timber extraction will be characterised by changes in forest biomass and tree size parameters, different tree architectures and tree species compositions. However, in some situations, more direct measures of human impact may be appropriate. The distance from a forest site to the nearest settlement or road may be correlated with the degree to which it has been disturbed or is used by people. Similarly, the number and width of paths in an area may be useful in characterising human disturbance, as may counting people directly. Which measures of human impact you choose will depend on the sort of impact thought to be important in the species studied.

6.3.3 Preparation of data for analysis
Ensuring accuracy
Estimation of habitat parameters saves time and energy, but measurements will always be more accurate than estimations. An alternative to estimating many parameters is to measure a few, though in this case you must try to ensure you will record the right habitat features. Many of these will be correlated with each other (see later), so for maximum effect for minimum effort, you might try to measure or carefully estimate around ten variables which are thought not to be closely autocorrelated.

As with estimating distances to bird contacts (Section 3), estimates of tree girths, distances to trees, percentage vegetation covers and other parameters are open to between-recorder errors. There may also be a tendency for recorders' estimates to drift during the fieldwork. Periods of habitat recording training prior to fieldwork and re-training during the fieldwork are essential if your habitat data are to be meaningful. With some parameters, such as vegetation cover, it may not be the actual value that is important, but how that value differs from the values of other sites.

If you think that there may be a problem with the accuracy of your habitat data, then 'rounding up' during the analysis phase might help. For example, ground vegetation covers, initially estimated to the nearest 5% could be lumped into 10% or even 20% bands. At its extreme, continuous data could be transformed to a simple one-zero format and perhaps some particularly inaccurate data points excluded from the analysis. Finally, if different recorders' estimates of, say, canopy cover are clearly different, then all values for a particular recorder could be scaled to the same mean as those of a different recorder.

Identifying autocorrelation
Many habitat features will be correlated with one another. For example, trees with large girths will tend to be tall, while forest with a full canopy may have a sparse ground cover (Figure 34). If habitat features are correlated with one another, then it may not be necessary for you to record all of the features. Identifying autocorrelation between your habitat variables can save you time in the field and will also simplify your habitat data set. During the data analysis phase, relating your bird data to relatively few habitat variables is usually better than relating them to many (as long as the few habitat features you have considered are appropriate). This is because the more different habitat features you consider, then the greater is the chance that you will

come up with 'spurious' correlations between your bird distribution and your habitat features. An alternative to collecting fewer habitat variables is to condense your many, inter-related habitat features onto just a few axes of habitat variability (section 6.4.5).

Figure 34. Correlation matrix of selected habitat variables (data from forests on Sumba, Indonesia). Figures shown are Pearson's Correlation coefficients. * $p<0.01$, ** $p<0.001$, NS: not significant.

	Canopy	Low	Ground	Average Girth	Above
Altitude	+0.21**	+0.21**	+0.09 NS	-0.14*	+0.27**
Canopy		-0.28**	-0.41**	+0.26**	+0.44**
Low			+0.49**	-0.11 NS	-0.24**
Ground				-0.20*	-0.33**
Average Girth					+0.09 NS

Data transformations

Habitat data will often need transforming prior to analysis. Values of some variables will have a positive skew (too many very large values). In these cases, values (x) should be log transformed ($\log(x)$), or if some values are zero in the original data, then $\log(x+1)$ should be used. If values of a habitat variable conform to a Poisson distribution, then square root transformation will make the distribution of values more symmetrical (or square root ($x+0.5$) if there are zeros in the data).

Many habitat variables will be collected as percentages (e.g. foliage profiles) or proportions (e.g. tree architectures). In these cases, arcsine transformation is usually performed. Computer programs will require you to transform the original data to proportions (between 0 and 1) before arcsine transformation.

Preparing your bird data

The sort of bird data you have will depend on the census method you used. Most commonly, you will be relating habitat features to the presence or absence of birds in a particular area. This might be variable circular plots at which you did and did not record a bird species, or sections of a transect, or small patches of forest in which a bird species sang and those in which it didn't. Each of your census points then becomes either a 'positive' (species recorded) or a 'negative' (species not recorded), and you look for differences, in terms of habitat, between the positives and negatives. An important consideration is the degree of confidence you have in your bird data - how sure are you that some of your negatives weren't in fact positives but you just failed to record the bird there? Clearly, the more times you have looked for a bird in an area and failed to record it, then the more certain you are that it is not there. You may want to only classify as 'positive' those plots at which you recorded the bird perched (i.e. not just flying over the area) and those at which the species was recorded at less than a certain distance away from the plot (e.g. 30–100m). Also, if you are not certain if the plot is positive or negative, then you may want to omit it from the analysis.

Because it is easier to prove presence than confirm absence, a better alternative is to use the abundance of the bird species at each of your sites. If for example you repeat-surveyed each of your census sites ten times, then you could express the abundance of a bird as the proportion of times out of ten that you recorded it at each site. It may be appropriate to combine bird data from several census plots together for analysis. All stations along one transect could be combined, as could all stations which share similar altitudes. Even better would be an estimate of bird density at each site, so long as estimates were precise enough.

6.4 Analytical approaches

6.4.1 Summary statistics

Although there is no doubt that all birds live in a multivariate world, it is still possible to obtain important information by studying the relationships between a species and a single habitat feature. Univariate and bivariate statistical tests should always precede more complex methods.

A simple table of means or median values for different habitat measures can contain a lot of information about habitat associations. Such tables can help in the interpretation of multivariate analyses. The inclusion of measures of dispersion, such as the standard deviation, is also helpful. Statistical tests such as Student's t or the Mann–Whitney U can be used to compare single habitat variables between sampling units with and without a species. If data are available from more than two areas, the habitat uses can be compared with an analysis of variance or an equivalent non-parametric method.

6.4.2 Indices

The index approach investigates if resources are used in proportion to their availability by calculating an index (Bookhout, 1994). Neu's method is commonly used (Figure 35).

Figure 35. Neu's selection index using simulated data.

Habitat	Availability	Usage		Index	
	Proportion a	*Records*	*Proportion r*	*Selection w*	*Standardised*
Primary forest	0.503	47	0.662	1.316	0.446
Secondary forest	0.220	21	0.296	1.344	0.455
Logged forest	0.145	3	0.042	0.291	0.099
Agriculture	0.133	0	0.000	0.000	0.000
Total	**1.000**	**71**	**1.000**	**2.951**	**1.000**

Calculations:

Selection index $w = r/a$ e.g. $0.662/0.503 = 1.316$

Standardised index $B = w/\Sigma w$ e.g. $1.316/2.951 = 0.446$

If the selection index >1 the habitat is preferred – usage is greater than availability. Standardising the indices allows comparison between studies because they always sum to one.

6.4.3 Graphical and linear regression approaches

If data have been collected from VCPs or transects, there will be quite a few data points. These can be ranked along a habitat gradient and grouped. Either the percentage of points occupied or the density estimate derived from DISTANCE can be plotted against a habitat gradient. Figure 36 shows the relationship between population densities of a parrot species and altitude in eight forest patches on an Indonesian island. In this case there is a negative, non-linear relationship between bird abundance and altitude. Such data are often tested by a regression analysis which models the relationship between bird abundance and habitat.

Figure 36. Relationship between cockatoo density and altitude.

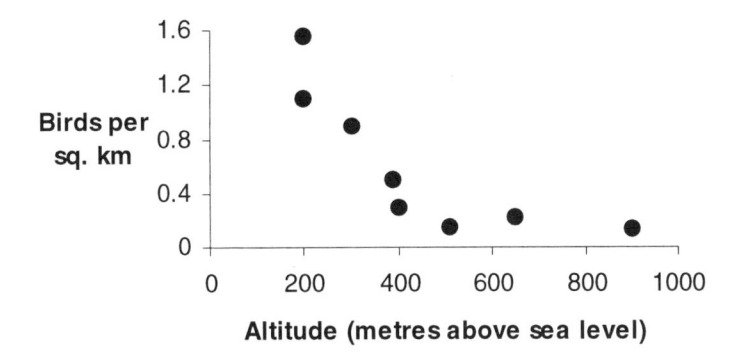

112 Expedition Field Techniques

A caution is appropriate here. There is a theoretical notion that species live among gradients of habitat. These will undoubtedly be of many dimensions, but imagine the simplest one dimensional example, say between abundance and altitude for the study bird (Figure 37). There is a preferred altitude, which is where the highest densities occur. Above and below this altitude, constraining factors come in – perhaps competition from a congener or maybe weather factors. At these altitudes, the bird is less common and eventually, some way from its optimum, it never occurs. If this situation was studied, the mean (or median) altitude and its dispersion would well describe the findings. A correlation of abundance with altitude would show no relationship. In another place, the forest available for study may be limited in altitude because the particular mountain is less high or the lowland forest has been felled. Here, the mean altitude is a biased estimate of the preferred altitude. On the other hand, there will be significant relationships between abundance and altitude (one positive and one negative) depending on the circumstances studied. Clearly these results depend on the circumstances in which they were measured and cannot be generalised.

Figure 37. The effect of survey area on bird-habitat correlations. The example shows a bird species which occurs at maximum abundance at around 600 m a.s.l. If study areas included only altitudes less than 600m, then there would be a positive correlation between bird frequency and altitude (A). The converse would be found if only altitudes above 600m were surveyed (B). Study design C gives no linear relationship.

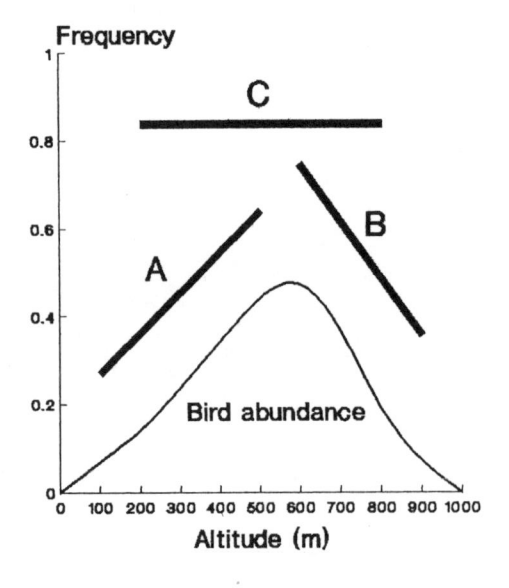

6.4.4 Logistic regression

If the data take the form of presence or absence at a series of plots (VCPs or transect stretches), a linear regression model is inappropriate, since each record can only take the value of 0 or 1, with nothing in between. Logistic regression is a powerful approach which can handle this problem, either for a single variable or for many (Jongman *et al.* 1995). There are ways to pare down the predictor variables to a small but effective (parsimonious) set. An attractive feature of the approach is that the model equation predicts the probability of the bird occurring at a point with any one set of habitat variables. So, within the normal constraints of regression models, you have an equation with the capacity to model effects of habitat changes. Logistic models might be developed for a number of species.

6.4.5 Reducing the dimensions

It is typical to measure many habitat variables which turn out to be correlated one with another. For instance, shrub cover might be inversely related to canopy cover (because scrub grows best where the canopy lets in more light). A canopy dwelling nectivore would be positively correlated with scrub cover but the relationship is deceptive. One way round this is to reduce the dimension of the habitat variables. Principal Components Analysis (PCA) is a common approach. The output is a small set of variables which are weighted sums of the original variables, explain a high proportion of the original variation and are independent of each other. Provided they can be interpreted in plain language, this can be quite a helpful approach. The difficulty is that it is possible to explain almost anything with enough imagination. PCA, logistic regression and other multivariate techniques are available in SPSS (Norušis 1993).

Correspondence Analysis or Detrended Correspondence Analysis are related to PCA. The latter is particularly popular in ecology and is implemented by a computer program DECORANA. DECORANA avoids some of the statistical assumptions of PCA which ecological data often violate. The habitat data set might be simplified by DECORANA before being entered into regression models.

It is possible to go one stage further and ask how bird communities might relate to habitat structure. In this case, a set of models for individual bird species is not very helpful. Canonical Correspondence Analysis (CCA) is a technique that identifies correlations between simplified bird and habitat axes. A program called CANOCO can implement CCA (ter Braak, 1987).

6.4.6 Interpreting and testing the results

The correct way to view these techniques is that they help with the exploration of data and the generation of hypotheses to explain what is happening. So, you may be able to say that a bird species is associated with low altitude, or that it occurs at higher population densities in forest areas with large trees than in logged areas. Such hypotheses should then be subjected to independent testing on new data sets. These data sets could be from a different area within the region, or in a different habitat. Several methods are available to test the validity of the habitat models you have formulated:

Resubstitution	The same data are used both for 'training' (formulating the model) and for 'testing' (in the new situation). Tends to overestimate predictive power.
Prospective sampling	The habitat model is developed from the original data, the model is tested on a new sample of cases (e.g. from a different area).
Data partitioning	The original data set is split up into a training and a test subset. The validity of the training model is tested on the test subset. Techniques include Bootstrapping and Jack-knife sampling.

Section 7
MAXIMISING THE IMPACT OF THE WORK

Colin Bibby

7.1 Basic communications

The political and economic issues surrounding biodiversity loss make the subject of resource conservation a sensitive one. Local people may be suspicious of outsiders. Government and people may have conflicting views as to who should benefit from natural resources such as timber and who should pay for the consequences of damage to the environment. Very commonly the benefits are taken by external groups and in a short term while the price is paid locally or by future generations. Protected areas may be perceived differently in a government office and on the ground. For these reasons, biologists from the capital or from abroad need to be sensitive to the fact that their aspirations for conservation may not be understood or shared by local people.

Communication is about dialogue. Maximising the impact of a study involves working out who the target audiences need to be and what messages are appropriate. Methods of delivery can then be considered. A simple scheme is illustrated in Figure 38. Note that some of the messages are written, some spoken and some communicated non-verbally, by behaviour and attitude. Scientists tend to think about the written word and its quality. The majority of the people directly involved in the environment do not. A technically excellent study might have limited or even negative impact on the ground if the scientists were perceived as rude, discourteous, or insensitive to local issues and culture.

Figure 38. A simple communications matrix for maximising the impact of a bird survey.

Audiences	Messages	Delivery
Local people	We are interested in this area because…..	Clear verbal
	Our interests are not a threat to you	Tactful and respectful attitude
		Openness

Regional and national technical (e.g. forestry department, protected area manager)	Here is some information that you may find useful	Good diplomacy Clear and simple written reporting which fits with their needs
Scientific – NGO or government; national or international	Here is a sound description of the status of a bird, a site or a habitat	Scientific publication Unpublished report Archive data

The underlying causes of threat to the world's birds are some very large issues (Figure 39). Amongst these, lack of knowledge and poor use of the information that is available is only a relatively small part. Knowledge about birds is again only a small part of the total gap in data and information required on ecology, politics and economics. In spite of this cautionary note, ornithologists have been major leaders in the conservation world for the simple reason that they have been good at exploiting the value of birds as indicators and communicating key information in an effective way.

Figure 39. Fundamental causes of biodiversity loss.

- Unsustainably high rates of human population growth and natural resource consumption.
- Steadily narrowing spectrum of traded products from agriculture and forestry, and introductions of exotic species associated with agriculture, forestry and fisheries.
- Economic systems and policies that fail to value the environment and its resources.
- Inequity in ownership and access to natural resources, including the benefits from use and conservation of biodiversity.
- Inadequate knowledge and inefficient use of information.
- Legal and institutional systems that promote unsustainable exploitation.

From the Global Biodiversity Strategy (WRI/IUCN/UNEP 1992)

Figure 38 could be considerably elaborated for a particular study. For the purpose of this section, its three major divisions of audience will suffice.

7.2 Culture, politics and diplomacy

Diplomacy is about being sensitive to circumstances which influence the prospects for effective dialogue.

Legal requirements
Complying with legal requirements can be quite hard work and may appear to be slow and bureaucratic. Ignoring legal requirements puts your own work in jeopardy but is also unhelpful to future relations between scientists and whoever's regulation it is that you have ignored. Things that might require permits over and above entry visas include travel, access to protected areas, research, mist-netting, collecting specimens (birds or anything else), or exporting specimens. To be effective, you need to find out what permits are required and how to get them – you may need letters of invitation or reference, photographs, or a preceding permit such as the right kind of visa. You will certainly need time.

Local involvement
If you are working in a distant place, in your own country or abroad, one essential step to good diplomacy is to work with local people. Students from the region will often appreciate the chance to join a survey and learn new things. You might employ guides or other local helpers. They will pay for themselves abundantly in their ability to speak the right language, interpret local nuances of behaviour or meaning, know how to go about things in the area, and maybe even know some of the birds. Few sponsors will want to support expeditions without local involvement for the simple reason that they are less likely to be successful. If you ask for help from local individuals or organisations remember that you might be involving them in costs. Internationally funded studies should certainly be sensitive to this and willing to defray such costs.

Cultural sensitivity
The risk of causing offence by oversight of cultural differences is ever present. The best way to deal with it is to have local participation who will be sensitive to the importance of everything from how you sit, how you dress, how you eat, where you wash and which sacred forest is a no-go area even if it is the oldest and best forest in the area. The only other rule is to behave at least as well as would be expected at home even if it is tempting to do otherwise while living in the forest. You will be ambassadors for conservation and your impact, for good or bad, will last longer than your visit.

Respect
You will never communicate anything to someone who feels that you have slighted them with disrespect. This might include failing to visit them and talk and keep them in touch with what you are doing. It takes time which could otherwise be spent in the field but if you want the results to have impact then you need your audience on your side.

Talking

Essentially good diplomacy comes down to forming good relations with people – being interested enough to talk with them and listen to them. Avoid being condescending but tell people what you are doing and why in plain language. If you can speak even a few words of the local language, you will come over as much more genuinely interested and involved. Local people may very well not know that the bird which occurs in their forest is very special and does not occur all over the world. You can give people something to be proud of by helping them to understand better the birds in their own area. At the same time, listen to what they have to say. Rural people may be very knowledgeable, but in a different way from that understood from a scientific training.

Figure 40. Five areas where good diplomacy will help your work to be influential. Poor diplomacy may override otherwise good work and thwart its influence.

- Legal requirements
- Local involvement
- Cultural sensitivity
- Respect
- Talking

7.3 Summary reporting

Field expeditions often produce two kinds of report. Commonly they produce their own report for limited circulation to helpers, sponsors or friends. They might later go on to publish something in the scientific literature. I will discuss these two separately though it might be a good idea to use the scientific paper as your summary report if you can write it quickly enough.

A few key considerations can increase the chances that your summary report will have the best chances of being effective.

Figure 41. Key points for the impact of your technical report

- Get it to the right people
- Produce it reasonably quickly
- Give people relevant information
- Summarise the key results
- Why are these results important?
- Keep the report brief
- Leave out irrelevant material
- Avoid naive political or economic recommendations

Get it to the right people
Good contacts and discussions should have helped in the process of identifying the key people from local communities, regional or national government departments and NGOs who might welcome your report. Some of them will already have been helpful in organising permits or other ways. These people will be looking forward to hearing about your findings, especially if you have communicated well with them during the trip. Remember to acknowledge and thank all your helpers and supporters. Consider translating it into the appropriate local language if this is different from yours – they will feel even better about the project when they see this.

Get it written by the right people
The further you are from home, the more important is the authorship. Local co-authors are the more likely to feel a share in ownership of the work if they have been given a chance to make a real contribution as co-authors. Local ownership is likely to make a big difference to the long term impact of a project.

Produce it reasonably quickly
If you take ages to produce a report you give the impression that it is not very important to you. How, in these circumstances can you expect your readers to think that your work is important. Some permits or grant-giving bodies may request a report within a given time period. Foreign visitors with research permits in some countries are expected to report before leaving. This need not be very difficult and is certainly a good discipline. Remember that all sorts of other things in life will take over once you have left and it might seem very difficult to write anything. It also gets harder to write once the fieldwork seems a distant memory. One way to increase the chances of writing a quick report is to be thinking about it while in the field. Keep track of the data you are collecting as they come in. Add up simple statistics as you go along such as the numbers of point counts completed in different habitats and the numbers of species recorded. Keeping all the data in one place rather than entrusting them to various people's field notebooks is another way to save time when it comes to sorting results out.

Give people relevant information
Remember who your audience is and provide them with information that they might find interesting and useful. This calls for some thought and understanding of the context of the recipients. What powers might they have to do something with the results? Which results do they most need to stimulate some action? Remember that many of your important readers will

not share your passion for conservation but will see it in relation to their lives and jobs.

Summarise the key results

Results are much easier to read if the author knows what the key points are and has made them clearly. A good discipline to help do this is to write the summary first as a series of single sentence factual points. There should be no more than about 10–15. Get these into a logical sequence and you then have not only a summary but also a synopsis of the report which will then almost write itself. Any material which is not logically needed to support the summary is of questionable value. The summary should come first – many readers will only have time for the first few pages.

Why are these results important?

Remember to think why your results are important and set sufficient background context to draw out their value. Point out that your data extend the known range of a globally threatened species. Indicate that you have found a very rich concentration of species in an Endemic Bird Area. Say that the data you have collected provide a baseline against which future changes of management in a protected area might be monitored. Be proud that you have trained a local student to be able to conduct further bird surveys in the region.

Keep the report brief

Short reports are harder to write than long ones but much easier to read. Many of your readers will be busy. Your report may not be in the language that they find easiest to read. It is a courtesy to readers to make the effort to express yourself briefly and clearly and it greatly enhances the chances of them reading and taking in what you have said. It is a mistake to believe that long and heavy reports are a sign of a weighty piece of work. I can think of no reason why anyone might prefer a long report to a brief and well written one – try to aim for a maximum length of ten pages. These comments apply to all possible readers including friends, funders or potential future employers whom you will want to impress. It is a courtesy and a help to local readers to provide at least the summary in the right language. Your local collaborators will presumably be able to translate if for you.

It may be that you have extensive scientific analysis to make and these will not be ready in time for a report to be read by your first target audience. Don't worry. The scientific report can come later. It will undoubtedly include some further new information. It may well be that you can already make many of the key points without the analysis – you can certainly report on the areas that the analysis will cover. You could say for instance that '*1695 bird records were collected at 120 point counts in six different habitats. Locally special (endemic) species appeared to be more frequent at higher altitudes but richness of species was greater in the lowlands. A further analysis will correlate species occurrence with habitat features'.* This will not prejudice your scientific report but will communicate much of what you have done and found to a key audience. There is a risk that your summary report will be quoted by other authors as if it was a scientific paper. Be scrupulously careful about what you say to avoid being misunderstood. Numbers and standard errors are a particular problem. If you have estimated a population as between 2000 and 10000 say so. 6000 ±4000 might be misquoted as 6000! And what about the units? Did you count pairs, singing males, or the whole population? The literature is full of erroneous quoting of pairs for individuals and vice versa.

Leave out irrelevant material

It should be obvious from the above that irrelevant material has no place in such a report. This is a difficult lesson. It is much easier to write pages of irrelevance than to produce a simple and sharp summary report. Think carefully about your audience before describing how you travelled, what you ate, what medicines you took, who lost their binoculars and sundry items that I have read in many a report from a field trip to an exciting place. Who wants to know about this sort of thing? Is it just yourselves? If so, the report is no place for it. Keep these topics for newspaper pieces or talks where they will go down well. If you must, then why not have a 200 page appendix to a brief report?

Avoid naïve political or economic recommendations

Unless you are very knowledgeable on the political and economic circumstances of the area you are studying, it may be ill advised to make any comments on these obviously sensitive subjects. Doing so risks annoying or offending the reader and may encourage them to suspect that if your politics are so naïve then perhaps your ornithology is as well. It is legitimate to say that the forests on such and such a mountain are being extensively cut and this is having an adverse effect on this threatened species whose population has become so small as to be at risk of extinction. To say that this practice must be stopped risks sounding naïve if you do not know about its legality or its economics (it may be the greatest source of revenue to the area). It may be legitimate to say that such and such a species is hunted by people in such and such a village. Again it may be naïve to say that this must be stopped – the relationship between the village and the management of the national park may be a matter of considerable political sensitivity. There are places where these issues can be addressed but a summary report of a field expedition is not one of them. Campaigning NGOs find it easier to base their arguments on scientific data if they can claim that the data are collected by sound scientists simply reporting facts. The local NGO can add the political spin it judges appropriate in advocating a particular action. If a scientific reports appears to contain questionable political or economic sentiments it is easy for officials to dismiss.

Make it look good
It may seem a shame but the appearance of a report is almost as important as its content when it comes to making an impression. With modern word processing it is not difficult to make a report look smartly designed and laid out. The time put into this will certainly give a better impression. Colour photographs can be reproduced quite cheaply but do not use them unless they say something worthwhile.

7.4 Scientific reporting

I do not know who first said that *'work unpublished is as good as work not done at all'*. Excepting the consideration that a summary report might be enough for some target audiences, this aphorism is largely true. The reason for publishing a scientific paper is that your information will potentially be available to ornithologists and conservationists the world over and for all time. A published paper means that the data and analysis have been refereed and also that they can be checked by someone doing subsequent work. Published papers are valuable, of course, for career prospects and personal satisfaction, if this is important to you.

A sensible precaution before writing a scientific paper is to choose the journal in which you would like to get it published. Read a few issues to see if they contain the sort of paper you expect to write. Read the guidelines for authors to check that your paper will be written within the scope of the journal and appropriate in length and style. Before you finish, pay scrupulous attention to the guidelines on such matters as citation of references, layout of legends, headings and other conventions. There is no greater way of irritating editors than to ignore their rules and guidelines. To do so suggests that your respect for the journal is low and irritated editors may be less sympathetic if they have any other reasons to hesitate over accepting your work.

There are many texts on writing scientific papers but not much sign from the literature that all authors have ever read any of them! Many of the points in the previous section are just as relevant to a scientific paper as to a summary report. Clarity of a paper depends on brevity, simplicity and good organisation. Writing a draft of the summary first is a good idea because it give focus to the important content of the paper which the rest needs to document and support. If the summary consists of single sentences each of which summarises something of the aim, method, finding or conclusion, so much the better. I find it helpful next to plan the tables and figures needed to document the evidence and present the analytical arguments and conclusions. Few journals will accept more than about ten tables and figures in total, but if they are the right ones, you can say a lot with ten. Major headings and a synopsis come next. After that it is easy! It should be pretty obvious what is needed to hang it all together.

Essentially your paper needs to say why your study was important, how you did it and what you found. You will have dealt with the purpose of the study in the planning stage (and having read Section 1). The methods section needs to deal with location, field methods, steps taken to minimise bias, and how the effort was designed and distributed. A key test, within the confines of brevity, is that the reader should have a good chance of going to the same area and conducting a comparable study 20 years from now. The results section needs a general overview of the data collected and portrayal of the analysis from which conclusions were drawn. Show your draft to other people for comment. Try it on scientists but also on the person nearest and dearest to you. He or she may not follow the scientific detail but they will tell you what needs improving with greater honesty than most other people might.

Figure 42. Key points for a scientific report

- **Summary**
  Write this first
  10–15 single sentence factual points
  Informative and interesting in its own right
  It is all that some people will read
- **Methods**
  Write this last
  Sufficient detail to be repeatable
  Emphasise sampling design and control of bias
  Keep it brief
- **Results**
  Up to 10 tables or figures to tell the story clearly
  Sound statistical analysis to support the conclusions
  Enough of the detailed findings summarised for future to communicate the shape of the basic information but editors will not allow masses of data
- **Discussion**
  Say why the study is important
  What are the most significant results?
  How do the findings fit with previous knowledge on the species or place or whatever?
  What further scientific work is required?

7.5 Archiving the data

You may well have far more detailed information than you can readily publish in a scientific paper but which is potentially very valuable for the future. This can be written as an archive which might contain detailed maps of study plots and census routs, original census data referring to mapped and dated transects or point counts. Most likely, these data will have been computerised and can be archived as a database or spreadsheet file but most safely will have a paper printout as well (or alone). You may also have a systematic list of miscellaneous bird observations which are potentially valuable but generally too bulky to be acceptable to a journal. The archive report will be weighty but it has a limited audience. Most people would prefer to see either a short report or a scientific paper and only rather few will want the full details.

Your archive data would be appreciated by the nearest relevant institutions, be they a university, protected area headquarters, local government office or NGO. There may be national conservation organisations, such as a BirdLife Partner who would value and could use such information. Many countries are currently working on establishing national biodiversity data centres often as partnership organisations hosted by government but supported by NGOs. If there is a national ornithological or bird conservation NGO this would be another place to leave an archive copy. Finally, BirdLife International maintains an extensive database and reference library of published and unpublished information on globally threatened species and important sites. This material is available to and much used by people planning further studies or seeking particular information on threatened species. Unpublished material held in BirdLife's library is cited in Birds to Watch, the global inventory of threatened birds. I hope the results of your study will contribute to the next edition. You can ensure this by publishing or by lodging unpublished archive material for the use of others.

Section 8
REFERENCES AND FURTHER READING

Allport, G., Ausden, M., Hayman, P.V., Robertson, P.A. and Wood, P.N. (1988) The Birds of Gola Forest, Sierra Leone. *ICBP Study Report 38*, Cambridge.

Ausden, M. and Wood, P.N. (1991) The Wildlife of the Western Area Forest, Sierra Leone. *Special Report to the Forestry Department, Sierra Leone*. ICBP/RSPB, Sandy, Bedfordshire, UK.

Bennun, L. and Waiyaki, E.M. (1993) Using timed species-counts to compare avifaunas in the Mau Forests, South West Kenya. *Proceedings of the VIII Pan-African Ornithological Congress 366*.

Bibby, C.J., Collar, N.J., Crosby, M.J., Heath, M.F., Imboden, C., Johnson, T.H., Long, A.J., Stattersfield, A.J. and Thirgood, S.J. (1992) *Putting Biodiversity on the Map: Priority Areas for Global Conservation*. International Council for Bird Preservation, Cambridge, UK.

Bibby, C.J., Burgess, N.D. and Hill, D.A. (1992) *Bird Census Techniques*. Academic Press, London.

Blake, J.G. (1992) Temporal variation in point counts of birds in a lowland wet forest in Costa Rica. *Condor 94*: 265–275.

Bookhout, T. A. (ed) (1994) *Research and Management Techniques for Wildlife and Habitats*. 5th Edition. Wildlife Society, USA.

Buckland, S.T., Anderson, D.R., Burnham, K.P. and Laake, J.L. (1993) *Distance Sampling: Estimating abundance of biological populations*. Chapman & Hall, London.

Collar, N.J., Crosby, M.J. and Stattersfield, A.J. (1994) *Birds to Watch 2, The World List of Threatened Birds*. BirdLife International, Cambridge, UK.

Collar, N.J., Gonzaga, L.P., Krabbe, N., Madroño Nieto, A., Naranjo, L.G., Parker, T.A. and Wege, D.C. (1992) *Threatened Birds of the Americas: The ICBP/IUCN Red Data Book*. International Council for Bird Preservation, Cambridge.

Collar, N.J. and Stuart, S.M. (1985) *Threatened Birds of Africa and Related Islands: The ICBP/IUCN Red Data Book*. International Council for Bird Preservation, Cambridge.

Evans, M.I. (1994) *Important Bird Areas in the Middle East*. BirdLife International, Cambridge, UK.

Grimmett, R.F.A. and Jones, T.A. (1989) Important Bird Areas in Europe. *Technical Publication No.9*. International Council for Bird Preservation, Cambridge, UK.

IUCN Species Survival Commission (1994) *IUCN Red List Categories*. Prepared by the 40th Meeting of the IUCN Council, Gland, Switzerland.

Jones, M.J., Linsley, M.D. and Marsden, S.J. (1995) Population sizes, status and habitat associations of the restricted-range bird species of Sumba, Indonesia. *Bird Conservation International* 5: 21-52.

Jongman, R.H.G., ter Braak, C.J.F. and Van Tongeren, O.F.R. (eds) (1995) *Data Analysis in Community and Landscape Ecology*. Cambridge University Press, Cambridge, UK.

Kapila, S. and Lyon, F. (1994) *Expedition Field Techniques: People Oriented Research*. Expedition Advisory Centre, London.

Laake, J.L., Buckland, S.T., Anderson, D.R. and Burnham, K.P. (1994) *DISTANCE User's guide Version 2.1*. Colorado Cooperative Fish and Wildlife Research Unit, Colorado State University, Fort Collins.

Mackinnon, J. and Phillips, K. (1993). *A Field Guide to the Birds of Sumatra, Java and Bali*. Oxford University Press, Oxford.

Marian, W.R., O'Meare, T.E. and Maehr, D.S. (1981). Use of playback in sampling elusive or secretive birds. *Studies in Avian Biology* 6: 81–85.

Norušis, M.J. (1993) *SPSS for Windows Professional Statistics 6.1*. SPSS Inc., Chicago.

Pomeroy, D. and Tengecho, B. (1986). Studies of birds in a semi-arid area of Kenya. III – the use of 'timed species-counts' for studying regional avifaunas. *Journal of Tropical Ecology* 2: 231–247.

Rabinowitz, D. (1981) Seven forms of rarity, in Synge, H. (ed) *Biological Aspects of Rare Plant Conservation*. Wiley, Chichester: 205–217.

Robertson, P.A. and Raminoarisoa, V. (1997) *Manuel de Formation Pratique en Ornithologie*. Projet ZICOMA, BirdLife International, Antananarivo.

Stattersfield, A.J., Crosby, M.J., Long, A.J. and Wege, D.C., (1998) *Endemic Bird Areas of the World*. BirdLife International, Cambridge, UK.

ter Braak, C.J.F. (1987) *CANOCO – a FORTRAN program for canonical community ordination by correspondence analysis, principal components analysis and redundancy analysis (version 2.1)*. Agricultural Mathematics Group, Wageningen, 95pp.

Torquebiau, E.F. (1986) Mosaic patterns in Dipterocarp rain forest in Indonesia and their implications for practical forestry. *Journal of Tropical Ecology* 2: 301-325.

Wege, D.C. and Long, A.J. (1995). *Key Areas for Threatened Birds in the Neotropics*. BirdLife International, Cambridge.

WRI, IUCN and UNEP (1992) *Global Biodiversity Strategy: Guidelines for action to save, study and use earth's biotic wealth, sustainably and equitably*. UNESCO, Paris.

Bird Surveys 127

Section 9
SAMPLE INPUT AND OUTPUT FILES FOR 'DISTANCE' PROGRAM

Data are entered into DISTANCE in the form of a syntax file. In this file, as well as entering your data set, you will need to enter several commands specifying the format that your data are in, i.e. whether you used line transects or VCP methodology, whether measurements are in metric or imperial units, etc. In addition, DISTANCE provides a number of optional commands that can be used to override the default program and tailor it to your particular data set/specifications. The commands you choose will vary depending on study design, methodology used to collect the data and what you need in the output.

To illustrate these points, we have included two basic examples of real syntax files that can be used as guidelines. You will of course want to tailor these files to your own particular needs and should consult the DISTANCE users' guide manual (Laake *et al.* 1994) for further information.

Sample input

The left hand column illustrates an actual syntax command file that can be adapted to meet your specific requirements. The right hand column is a brief explanation of what the commands mean and why they were chosen. Unless otherwise stated, the commands used in the two examples are not specific to either line transects or point counts but can be used interchangably. It is important that the correct punctuation is followed throughout the input file, as the DISTANCE program will not recognise your commands otherwise.

Example one is taken from a line transect study of hornbills in Zambia (the data set is not complete). In this example, data were collected in two types of woodland (chipya and miombo) and several transects were conducted in each. The habitat types have been allocated as `Stratum` and the transects as `Samples` within these. This enables the calculation of a density estimate for each transect and habitat type individually, as well as a pooled estimate to be made across habitat types, if appropriate.

Options:	
`Title='HORNBILLS IN ZAMBIAN WOODLAND';`	* A title may be displayed.
`Distance=Perp/Exact;`	* Specifies that the distance data entered is of the perpendicular and exact format.
`Select=All;`	* All distance data is to be used in the analysis.
`Distance/Units=metres;`	* Instructs that the units of measurement for the distance data, entered under Sample, are in metres.
`Length/Units=Kilometres;`	* Instructs that the units of measurement for the length of transect, entered under Effort, are in kilometres.
`Area/Units=Square Kilometres;`	* Instructs that the units of measurement for the area of habitat, specified under Stratum, is in square kilometres.
`End;`	* Remember to include the End; command after each section.
Data:	
`Stratum/Area=44.6/Label= 'Chipya Habitat';`	* Habitat data have been split into two Strata and each assigned a label. If the area of habitat under study is known, a population estimate may be calculated, using the density estimates produced, by inserting the Area command.

`Sample/Effort=28.0/Label ='Transect1';`
 	
`46,3,50,9,42,34,30,2,4,8;`
 	
`Sample/Effort=36.0/Label ='Transect2';`
	* Each transect has been assigned as a Sample and labelled correspondingly. The Effort, i.e. the length of the transect multiplied by the number of repeats, is also indicated.

`24,48,50,2,140,25,27,53,`
`15;`
 	
`Sample/Effort=3.4/Label=`
`'Transect3';`
 	
`0,17,90,57,6;`	

`Stratum/Area=43.5/Label=`
`'Miombo Habitat';`
 	
`Sample/Effort=29.5/Label`
`='Transect4';`
 	
`2,18,7,150,19,14,5,2,18,`
`8,0;`
 	
`Sample/Effort=4.8/Label=`
`'Transect5';`
 	
`;*`
 	
`Sample/Effort=74.0/Label`
`='Transect6';`
 	
`70,25,10,11,15,17,15;`
 	
`Sample/Effort=21.6/Label`
`='Transect7';`
 	
`3.5,20,17,12,10,3,80,13;`
	* Note no observations were made on transect 5 but it must still be included in the analysis.

`End;`	
`Estimate:`	* The Estimate section determines which criteria you wish the density estimate to be calculated by.
`Density by stratum/`	* This command instructs for an estimate of density to be calculated for each habitat type (stratum). To obtain an estimate of density for each transect substitute with 'sample'. A pooled estimate of density will automatically be provided by the default.

`Weight by effort;`	* The contribution made by each transect to the density estimate for each stratum, is to be weighted by the length of that transect.
`Detection All;`	* If the study species has an equal chance of being detected in each habitat type (see ESW), then all distances can be used together (giving a larger sample size), when calculating the detection distance.
`End;`	

The second example is the basis for a VCP method study in cloud forest habitat, Ecuador.

`Options:`	
`Type=Point;`	* If point counts were used then this must be specified as the default setting is for line transects.
`Object=Cluster;`	* This command will allow estimates for observations recorded as clusters.
`Select=All;`	
`Distance/Measure=Metres/r truncate=0.20;`	* `rtruncate=0.20` instructs DISTANCE to truncate the data at the right hand side of the detection curve. This value can be altered and recommendations have been made in section 3.5.2.

`Dist/int=6,12,18,24,36;`	* This command can be used to instruct DISTANCE to allocate the distances measured into certain intervals. This is a group remedy to overcome 'heaping' (see section 3.4.2).
`Area/Units=Square Kilometres;`	
`End;`	
`Data:`	

`Sample/Effort=11.2;`
`34, 3, 18, 6, 2.5, 2, 57,`
`2, 9, 3;`
`Sample/Effort=5.4;`
`0, 1, 6, 4;`
	* Distances should be entered followed by the cluster size and separated by a comma. Cluster sizes are highlighted in bold.

`End;`	

`Estimate:`
`Density by sample;`
`Estimator/Key=Hazard;`
`Estimator/Key=Uniform;`
`Pick=AIC;`
 	
`End;`
	* Detection is to be estimated by sample, selecting between the two models without estimating further parameters. See section 3.5.2 for guidelines on model selection.
	* Instructs DISTANCE to choose the model with the lowest AIC value.

Sample output

DISTANCE output begins with a summary of the input data including: the number of samples, observations and strip width. Check here for any obvious errors. This is followed by the choice of the best model, with appropriate adjustments, to fit the data. After various test statistics, the output finishes with a summary of the final density estimates. Below is a brief outline of the major sections which will be useful to the interpretation of your density estimates.

`Density Estimation Results` – look here to see which model and adjustments have been finally chosen by the programme, using the AIC values (see below).

The chosen model is illustrated by a graph of a `detection probability curve` and a `CHI-SQ` goodness of fitness test is also performed. If a reliable chi-square test cannot be achieved, DISTANCE will call for you to pool some of your data by hand. This may mean adding data from separate transects or habitat types together to obtain larger sample sizes within the group.

This section contains several important statistics. In particular check:

ESW – Effective Strip Width is the distance beyond which as many observations are missed as are included within the strip (EDR in point counts). This value may differ between habitats and species, and can be used to decide whether two sets of data are suitable for pooling by hand.

AIC – Akaike's Information Criterion is used in model selection; generally the model with the lowest value is chosen for fitting.

Chi-p – Probability for χ^2 goodness of fitness test.

	Estimate	%CV	dF	95% Confidence Interval	
Stratum: 'HABITAT CHIPYA'					
Half-normal/Cosine					
m	3.0000				
AIC	519.12				
Chi-p	.29314				
f(0)	13960E-01	18.75	48	.96070E-02	.20286E-01
p	.23877	18.75	48	.16432	.34697
ESW	71.632	18.75	48	49.295	104.09

The last section of output presents the final density estimates.

	Estimate	%CV	dF	95% Confidence Interval	
Stratum: 'HABITAT CHIPYA'					
Half-normal/Cosine					
DS	5.2197	25.84	4	2.5770	10.572
D	7.9515	26.80	5	5.0403	15.649
Stratum: 'HABITAT MIOMBO'					
Half-normal/Cosine					
DS	16.310	34.25	23	8.18.76	32.488
D	26.529	35.36	27	13.118	53.654
Pooled Estimates					
DS	8.5682	22.55	25	5.4160	13.555
D	13.561	23.59	31	8.4354	21.801

Key points are:

DS – the density of clusters (if appropriate).

D – the density of individual animals per unit of measurement specified.

95% confidence interval – indicates that there is 95% probability that the density estimate falls between the two values specified. These values are useful for calculating maximum/minimum population sizes within specified areas.

The website for DISTANCE includes software that is free to download. The address is:

http://www.mbr.nbs.gov/software.html

The BP Conservation Programme

The BP Conservation Programme, organised by BirdLife International, the British Petroleum Company plc (BP) and Fauna & Flora International (FFI), aims to encourage long term conservation projects which address global priorities at a local level. Each year the Programme gives out advice, training and financial awards to teams of students all over the world building projects which fulfil the following criteria:

- address a conservation priority of global importance;
- have a strong association with the country where the project will take place (local people participating in all stages of the project);
- the majority of the team must be university students (under or post-graduates in full- or part-time study).

These specific criteria aim to increase the long-term, sustainable conservation achievements of a project by focusing the research objectives and building vital links between personnel at all levels, from project team members and local people to government staff.

Further information about this Programme is available on the web:

http://www.bp.com/conservation/

or from:

Programme Manager
BirdLife International/FFI
Wellbrook Court
Girton Road
Cambridge CB3 ONA
UK

Tel: +44 1223 277318
Fax: +44 1223 277200
Email: bp-conservation-programme@birdlife.org.uk